T0122989

DSM

DSM

A History of
Psychiatry's Bible

ALLAN V. HORWITZ

Johns Hopkins University Press
Baltimore

© 2021 Johns Hopkins University Press
All rights reserved. Published 2021
Printed in the United States of America on acid-free paper

9 8 7 6 5 4 3 2

Johns Hopkins University Press
2715 North Charles Street
Baltimore, Maryland 21218-4363
www.press.jhu.edu

Library of Congress Cataloging-in-Publication Data

Names: Horwitz, Allan V., author.
Title: DSM : a history of psychiatry's bible / Allan V. Horwitz.
Description: Baltimore : Johns Hopkins University Press, [2021] |
    Includes bibliographical references and index.
Identifiers: LCCN 2020030744 | ISBN 9781421440699 (hardcover ; alk. paper) |
    ISBN 9781421440705 (ebook)
Subjects: MESH: Diagnostic and statistical manual of mental disorders. |
    Mental Disorders—classification | Mental Disorders—diagnosis |
    Psychiatry—history | History, 20th Century | History, 21st Century |
    United States
Classification: LCC RC454 | NLM WM 15 | DDC 616.89—dc23
LC record available at https://lccn.loc.gov/2020030744

A catalog record for this book is available from the British Library.

*Special discounts are available for bulk purchases of this book. For more
information, please contact Special Sales at specialsales@jh.edu.*

To my daughters, sons-in-law, and grandchildren

Contents

Preface

The *Diagnostic and Statistical Manual of Mental Disorders* (*DSM*) of the American Psychiatric Association pervades our culture. Since publication of the manual's third edition in 1980, its diagnoses define what mental disorders are considered legitimate, how patients conceive of their problems, who receives government benefits, and which conditions psychotropic drugs target and insurance companies will pay to treat. They also delineate the curriculum that is taught to psychiatrists and other mental health professionals, the diagnoses that researchers and epidemiologists explore, and the psychic problems that public policies attempt to remedy. Conversely, the *DSM* has attracted an enormous amount of criticism from, among others, feminists, gay activists, the anti-psychiatry movement, academics from various disciplines, and many psychiatrists. No other medical manual has been the object of so much attention and critique.

This book traces the evolution of the *DSM* from its predecessors that arose in the nineteenth and first half of the twentieth centuries, to the emergence of the first analytically oriented manuals in 1952 and 1968, and to the incarnation of the medical model that has marked each *DSM* since 1980. It tells a tale of how diagnostic criteria, which must appear to portray evidence-based empirical research, in fact emerge from uncertainty, factionalism, intense political conflicts, economic considerations, and other interests. As these factors have evolved over the course of the past seven decades, the manual's content has accordingly altered.

When Matt McAdam, my editor at Johns Hopkins University Press, asked whether I'd be interested in writing a history of the *DSM*, my immediate reaction was "there's so many that have been done already, what's the point of another one?" After I gave more thought to his suggestion, however, I realized I was mistaken. Certainly, a large and insightful literature exists on some editions of the *DSM*. Hannah Decker's *The Making of the DSM-III* and Stuart Kirk and Herb Kutchins's *The Selling of DSM* provide compelling, and contrasting, accounts of the development of the manual's third edition. More recently, Allen Frances's *Saving Normal*, Gary Greenberg's *The Book of Woe*, Edward Shorter's *What Psychiatry Left*

Out of the DSM-5, and Rachel Cooper's *Diagnosing the Diagnostic and Statistical Manual of Mental Disorders* do the same for the latest version, the *DSM-5* (2013). Jason Schnittker's *The Diagnostic System* provides a penetrating analysis of the different ways that various interests use the *DSM*. Yet no single volume has considered the manual's evolution from its first edition in 1952 through the present. This book attempts to fill the resulting gap.

In addition to these books, a voluminous literature explores various aspects of the *DSM*. I have especially benefited from the writings of Anne Figert, Michael First, Gerald Grob, Nick Haslam, Kenneth Kendler, Christopher Lane, Richard McNally, Joel Paris, Jerome Wakefield, Owen Whooley, and Peter Zachar, to name a few influences. In addition, this book stands on the shoulders of the many scholars who have produced major studies about the broader social and historical context of modern American psychiatry. They include Ronald Bayer, Anne Harrington, David Healy, Ellen Herman, David Herzberg, Elizabeth Lunbeck, George Makari, Marc Micale, Jon Metzl, Richard Noll, Charles Rosenberg, Andrew Scull, Robert Spitzer, Michael Staub, and Andrea Tone. None of them is responsible for the ways in which I have used their work.

I am particularly indebted to Christopher Lane, who generously provided me with his files on the correspondence about the development of the *DSM-III* and *DSM-III-R* that are found in the Melvin Sabshin Library and Archive of the American Psychiatric Association Foundation in Washington, DC. I am also grateful to Deena Gorland, the librarian at the archive for granting me permission to use this material. David Herzberg graciously sent me the data on psychotropic drug use in the 1950s and 1960s that he extracted from the National Disease and Therapeutic Index. David Mechanic's comments on an earlier draft of the manuscript greatly improved the final product. Michael First and Peter Zachar kindly responded to my queries about the *DSM* process. Matt McAdam, Will Krause, and Kyle Kretzer at Johns Hopkins University Press provided invaluable support. Finally, Carrie Watterson was a superb copy editor.

DSM

ONE | Diagnosing Mental Illness

Diagnoses from the *Diagnostic and Statistical Manual of Mental Disorders* (*DSM*) have become part of our culture. Environmental activist Greta Thunberg is widely known to have Asperger syndrome; one of her adversaries, Donald Trump, is commonly seen as displaying a narcissistic personality disorder. The singer Mariah Carey has discussed her struggles with bipolar disorder, the same condition dramatized in the main character in *Homeland*, Carrie Mathison. Lady Gaga has spoken about her struggles with posttraumatic stress disorder (PTSD). TV mobster Tony Soprano seeks treatment for panic attacks, while his son is suspected of having attention deficit / hyperactivity disorder (ADHD). Susanna Kaysen, the author of the best-selling memoir *Girl Interrupted*, discovered she had borderline personality disorder while reading the *DSM-III* in her local bookstore. Many patients enter therapy already knowing what diagnosis they expect to receive.

The centrality of *DSM* diagnoses is a new phenomenon, arising only after the third edition of the manual was published in 1980. Although psychiatry had been a pervasive presence in American culture since World War II, specific mental disorders were not. Instead, as a *Life* magazine story summarized in 1957, the field "permeates our whole way of life—at work, in love, in sickness, and in health."[1] The most prominent conceptions of mental illness at the time were broad notions such as "psychoses," "neuroses," and "nerves" or psychoanalytic processes such as

Oedipal or inferiority complexes, repression, and the id, ego, and superego. For example, the first sentence of Saul Bellow's *Herzog* (1964) reads, "If I am out of my mind, it's all right with me." Similarly, gang members in *West Side Story* (1957) sing about one of their associates:

> It's just his neurosis that oughta be curbed
> He's psychologically disturbed!
> We're disturbed, we're disturbed
> We're the most disturbed, like we're psychologic'ly disturbed[2]

In the same vein, tranquilizing drugs such as Miltown and Valium were promoted in the 1950s and 1960s for "the patient with vague symptoms, nervous and distressed under the burden of unsolved problems, [who] finds release from anxiety and restoration of emotional composure."[3] Specific diagnoses as we know them today had little importance in either professional practice or cultural consciousness.

Before the 1980s, psychiatric diagnoses were inextricable from the experiences of particular individuals. Clinicians viewed mental problems as being closely intertwined with people's psychosocial backgrounds and circumstances and therefore impossible to isolate from personal interpretations, identities, socialization, relationships, and life events. This view of mental illness sharply divided psychiatry from other medical specialties, which studied diseases with predictable courses and outcomes that were independent of particular individual lives.

Since the *DSM-III*, however, psychiatric diagnoses have been considered comparable to organic diseases that "can and should be thought of as entities existing outside the unique manifestations of illness in particular men and women."[4] Mental, as much as physical, diseases are discrete ailments with characteristic causes, prognoses, and outcomes. Their diagnoses stand apart from the singularity of the individual patient. "A common misconception is that a classification of mental disorders classifies individuals, when actually what are being classified are disorders that individuals have," the *DSM-III* states.[5]

The history of the *DSM* illustrates the movement from conceptualizing mental disorders as aspects of individual experiences to viewing them as features of common disease processes that characterize everyone who has them. A central issue in the manual's evolution is the extent to which mental disorders are comparable to those in general medicine or whether the intimate links of psychiatric conditions to selves, identities, symbolic processes, and cultures render their classifications fundamentally different from those of organic diseases.

The *DSM*

The *Diagnostic and Statistical Manual of Mental Disorders* of the American Psychiatric Association (APA) provides the gold standard for psychiatric diagnoses. First published in 1952, the manual is now in its fifth edition, *DSM-5* (2013). Its first two versions, the *DSM-I* (1952) and *DSM-II* (1968), presented brief descriptions of the major mental disorders but were obscure in both the general culture and within the psychiatric profession itself. The manual first became publicly visible in the early 1970s when the *DSM-II* classification of homosexuality as a mental disorder created a highly publicized uproar. The *DSM* gained prominence only in 1980, after the publication of its third edition, which became the heart of professional education, practice, and research. Since that time it has been the reference point for all mental health professionals, even those who disdain it. The manual establishes which psychiatric conditions are taught in medical and other professional schools, determine eligibility for disability payments for patients and insurance compensation for providers, are targeted by pharmaceutical advertisements, become objects of psychiatric research, and shape public formulations of mental illness. It is also firmly embedded in the administrative apparatus of hospitals, private practices, the judicial system, and all other institutions that deal with mental disorder. Moreover, the *DSM* shapes the way individuals conceive of their own psychological problems.

The importance of the *DSM* for psychiatry is unique among medical specialties. Other areas of medicine commonly rely on biological markers that confirm or refute a diagnosis of some disease: cardiologists use PET scans to see whether a heart has tissue damage, nephrologists take x-rays to search for a kidney stone, oncologists perform biopsies to detect cancerous cells, and general practitioners employ blood tests to establish levels of cholesterol or blood pressure. Psychiatrists, however, have none of these tools. The lack of confirming markers for any common mental disorder means that diagnosis in itself has an outsize role in psychiatry compared to other branches of medicine. Although most of the *DSM* diagnoses lay out detailed symptom criteria and specific inclusion and exclusion rules, in actuality, patient self-reports and, sometimes, clinical observations constitute psychiatry's diagnostic resources. No independent criteria can verify the accuracy of a clinician's assessment of a mental disorder.[6]

The American Psychiatric Association owns the *DSM*, which allows the organization to monopolize the diagnoses of mental disorder for all mental health professions. The American Psychoanalytic Association published its *Psychodynamic*

Diagnostic Manual in 2006, but it is rarely used, culturally invisible, and does not authorize insurance reimbursement. Other assessment tools such as the Myers-Briggs Personality Inventory (MBPI) are well known and widely applied but do not identify mental disorders. The MBPI, for example, categorizes everyone as having 1 of 32 personality types that are all considered normal; none indicates pathology. The World Health Organization's *International Statistical Classification of Diseases and Related Health Problems (ICD)* is similar in many ways to the *DSM* but has little presence in the US health care system. Even worldwide, the *DSM* has come to be more commonly used than the *ICD*.[7] The National Institute of Mental Health (NIMH) is developing a new diagnostic system, the research domain criteria (RDoC), which is solely intended for research, not clinical, purposes. The RDoC, however, is not yet operational.

Its unchallenged position in the United States and, increasingly, globally has led the *DSM* to be called the bible of psychiatry. In fact, it is anything but inerrant scripture. The manual has undergone frequent and, sometimes, radical changes since its initial appearance in 1952. Heated debates, intense internal divisions, and sharply conflicting proposals marked the most recent revision process. Aside from using the *DSM* for rhetorical and institutional purposes, few psychiatrists consider its diagnoses accurate portrayals of underlying natural phenomena.[8] They do, however, religiously use them for educational training, obtaining reimbursement for treatment, submitting grant applications, providing measures for epidemiological studies, and all other activities where some diagnosis is necessary.

The *DSM* has become invaluable to the APA itself as well as to its members. Since the third edition, the organization has sold around a million copies of each volume, as well as assorted workbooks, sourcebooks, and diagnostic guides. There were 150,000 preorders for the *DSM-5*, which was initially priced at $199 for the hardback and $149 for the paperback edition.[9] These various publications are a major source of the APA's income. The *DSM* also allows psychiatry to control the diagnostic system that all mental health professionals employ. Clinical psychologists, psychiatric social workers, psychiatric nurses, mental health counselors, and other therapists must use the *DSM* categories to receive third-party payment for their services.

In addition, the manual serves as the benchmark for determining mental disorder in the judicial system. As of 2011, more than 5,500 court opinions cited the *DSM*. Its diagnoses are invoked in widely disparate legal areas including providing defense from criminal responsibility, exemptions from the death penalty, eligibility for disability benefits, and determinations in child custody cases. Legislative bodies, too, typically invoke the *DSM* in statutes that deal with definitions of mental disorder.[10]

The *DSM* not only controls what diagnoses treated patients receive but also defines mental disorders in the general population. The best-known research finds that at some point in their lives about 31 percent of people experience an anxiety disorder, 21 percent a mood disorder, 25 percent some impulse control disorder, and 35 percent a substance use disorder.[11] Other studies provide much higher appraisals, one finding that up to two-thirds of community members have suffered from just the single condition of major depressive disorder.[12] Moreover, rates of *DSM* disorders appear to be rising at an alarming pace: studies conducted in the 1980s, 1990s, and 2000s show continuously growing numbers of individuals with these conditions.[13] The use of *DSM* diagnoses thus makes it seem as if mental disorders are rampant in the population. Far from being a specialty that treats a small group of seriously disturbed people, psychiatry (and other mental health professions) is charged with a mission to confront a large and growing "public health epidemic" that threatens virtually everyone.

Without exaggeration, a former president of the APA claims that "the *DSM* might just be the most influential book written in the past century."[14] What is the nature of this diagnostic manual that has become the essential tool for defining mental disorders?

Three Interpretations of the *DSM*

Commentators have taken three general approaches to the *DSM*. The first views each edition of the manual since 1980 as an example of a steady, if uneven, growth of knowledge about the nature of mental illness. A second perspective regards the *DSM* diagnoses as stigmatizing labels imposed on patients. The third outlook sees the *DSM* as a product of intra-professional and external social forces. These interpretations need not be mutually exclusive: different observers can place more or less emphasis on how the manual embodies scientific progress, functions as an instrument of social control, or is the result of social factors.

The Embodiment of Scientific Progress

Since its third edition, many psychiatrists have viewed the *DSM* as the product of improving empirical knowledge about mental disorder. They regard the first two editions as reflecting the unscientific or even anti-scientific views of eclectic American psychiatrist Adolf Meyer and psychoanalyst Sigmund Freud. After the *DSM-III* revolution in 1980, however, scientific evidence has served as the foundation of psychiatric diagnoses. Therefore, the evolution of the manual represents the triumph of "science over ideology": "The old psychiatry derives from theory, the new psychiatry from fact." This progress has been most apparent over the past 40 years, as new research has allowed psychiatry to "mature from a psychoanalytic cult of shrinks into a scientific medicine of the brain."[15]

From this perspective, the *DSM*'s progress stems from the realization that psychiatric diagnoses are in essence equivalent to diagnoses in other medical specialties. The *DSM* criteria specify disease entities that can be identified and treated in similar ways regardless of the types of people who have them or the cultures in which they arise. Like other medical fields, psychiatry strives to develop accurate diagnoses that can predict the causes, prognoses, outcomes, and optimal treatments of particular diseases. Its conditions emerge in comparable ways as they do in any area of medicine—through the findings of procedures such as double-blind control studies, experimental investigations, and epidemiological research. As these techniques uncover new evidence or challenge existing definitions, diagnostic criteria accordingly will change for the better.

Harmful Impacts

A second view inverts the progressive narrative of the *DSM*. "Few branches of the medical profession have been as subject to complaints as psychiatry," historical sociologist Andrew Scull observes.[16] Its many critics regard psychiatry's claims about the scientific basis of the *DSM* as inflated at best and self-serving and hypocritical at worst. Since the 1960s they have pursued several lines of attack on psychiatric diagnoses, arguing, for example, that mental disorders cannot be described in the same ways as physical diseases, that the purpose of diagnosis is to control undesirable behaviors, and that the manual extends the realm of psychic pathology in unwarranted ways.

One common theme is that mental disorders are inherently unsuited to the molds that might fit the specific diseases entities of physical medicine. Most famously, in 1961 Thomas Szasz criticized all conceptions of mental illness as "myths."[17] Detractors like Szasz object that mental phenomena, which involve social values, consciousness, and huge variation across individuals and cultures, preclude the construction of any accurate diagnostic system. More recently, 29 British psychiatrists argued in a 2012 editorial that the *DSM-5* fundamentally mischaracterizes the essence of mental disorder: "Psychiatry is not neurology; it is not a medicine of the brain. Although mental health problems undoubtedly have a biological dimension, in their very nature they reach beyond the brain to involve social, cultural and psychological dimensions. These cannot always be grasped through the epistemology of biomedicine."[18]

In this view, other medical specialties typically deal with diseases that are distinct from, not continuous with, health. Psychiatry, however, encounters phenomena that often cannot be easily separated from conventional behaviors. Indeed, the field is concerned with conditions that display no sharp distinctions between pathology and normality. Therefore, diagnostic manuals like the *DSM* are inherently incompatible with the qualities of psychological disorders.

Another critique of this type focuses on how psychiatry is more likely to use *DSM* diagnoses for social control than for scientific purposes. The title of feminist psychologist Paula Caplan's book, *They Say You're Crazy: How the World's Most Powerful Psychiatrists Decide Who's Normal*, conveys the essence of this account.[19] The *DSM* diagnoses usually move from "powerful psychiatrists" to relatively powerless groups such as women, racial and ethnic minorities, the sexually unconventional, and the impoverished. Through placing medical-sounding labels on problems of living, they mask the true sources of distress, which often lie in inequitable and oppressive social circumstances. For these critics, what is called "madness" is more a consequence of civilization than its antithesis.[20]

A related variant of this critique focuses on how the *DSM* increasingly applies medical labels to normal emotions, exposing more people to the negative consequences that often follow from psychiatric diagnoses, such as stigma and the harmful side effects of drug regimens. The publication of the *DSM-5* led Allen Frances, the chair of the DSM-IV Task Force, to observe, "Many millions of people with normal grief, gluttony, distractibility, worries, reactions to stress, the temper tantrums of childhood, the forgetting of old age, and 'behavioral addictions' will soon be mislabeled as psychiatrically sick and given inappropriate treatment."[21] Each of these critiques regards the *DSM* as a document that is more likely to harm than to help those whom it calls "mentally disordered."

A SOCIAL VIEW

The social perspective provides a third way of thinking about the *DSM*. It suggests that the manual's diagnoses neither identify facts about the natural world nor are inherently coercive. Classifying things according to similarity to or difference from each other is an essential aspect of how humans organize their world. Mental illness arises because cultural and historical forces structure mental phenomena into diverse classifications that conform to social realities. Each *DSM* is a product of a psychiatric thought community that shares common assumptions at particular points in history. "Perhaps most fundamentally," historian Charles Rosenberg explains, "the act of diagnosis links the individual to the social system."[22]

Thus, diagnosticians can potentially classify symptoms that are considered to be signs of some "mental disorder" in innumerable ways. In 1857 two preeminent British psychiatrists explained:

> It is needful that [the student] should never forget, that convenient and necessary as are classifications and divisions, for the purposes of facilitating the comprehension of the multiform phases of insanity, which, without them, would present a more rude and undigested heap than is at present the case, Nature herself cannot be so precisely limited; and that, in her book, as opened to him in the wards of an asylum, he must be prepared to find a combination, a blending, if not a confusion of the elementary forms.[23]

The groupings psychiatrists choose to classify this "undigested heap" are not idiosyncratic but reflect socially organized processes. As sociohistorical circumstances change, diagnostic systems accordingly evolve.

To say that the *DSM* reflects the social matrix that surrounds it does not mean that the manual is nothing but a product of these conditions. Nor does it mean that any diagnostic manual is as good as any other. Instead, the social perspective regards psychiatric diagnoses not as simply mirroring of psychobiological phenomena but also as emerging from historical circumstances, cultural beliefs, and political and economic conflicts.

A variety of factors shapes what kind of diagnostic system dominates any particular era. Some stem from intra-professional sources. The *DSM* is a product of the American Psychiatric Association, which is responsible for upholding the legitimacy of the profession. During the 1950s and 1960s, analytically oriented psychiatrists, who scorned the use of specific diagnoses organized by observable symptoms, dominated the organization. Consequently, the first two *DSMs* paid little attention to developing precise specifications of diagnostic criteria.

After the 1960s, however, intense pressures developed from, among other sources, federal regulators, insurance companies, and medical schools to portray psychiatrists as doctors practicing medicine. In recent decades, their legitimacy stems from how they name, define, and distinguish their central concepts from each other: "Diagnosis is the first step in the technological process of transforming a person with an ambiguous complaint into a client with a defined mental disorder."[24] The credibility of the *DSM* now depends on its depiction as the evidence-based result of scientific research. This means that diagnoses must be believed to stem from empirically derived data, despite the evidence justifying many diverse interpretations of symptoms.

There are also clear intra-professional divisions regarding the role of diagnoses. In particular, researchers and clinicians use the *DSM* in distinct ways. Researchers require specific diagnoses to create homogeneous groups that can reveal the etiology, prognosis, and best treatments for the particular condition under study. Their definitions must be similar to those of others who study comparable phenomena. In contrast, clinicians must deal with the idiosyncrasies of particular clients. For them, diagnoses are practical tools, not the basis for standardized protocols. The continuing battles between clinicians and researchers have shaped recent editions.

A variety of other forces also influences the *DSM*. One is the National Institute of Mental Health, which partnered with the APA in shaping every *DSM* before a sharp break between the organizations arose during the development of the *DSM-5*. Since the 1960s, private and public insurance programs have also had major impacts on the *DSM*. These third parties set the parameters for which diagnoses are acceptable for reimbursement for treatment. In addition, patients require *DSM* diagnoses to obtain insurance coverage for drugs, psychotherapies, and other benefits. Furthermore, a variety of advocacy groups have formed to oppose the narrowing, promote the broadening, or, more rarely, abolish diagnostic criteria for particular *DSM* conditions.

Finally, pharmaceutical companies have been intimately connected to diagnostic classification systems. Since the early 1970s, the Food and Drug Administration's (FDA) regulations have required the drug industry to market its products as treatments for particular *DSM* diagnoses. Drug companies are also a major source of income for departments of psychiatry in medical schools, psychiatric researchers, and the APA. The web of affiliations between the industry and the psychiatric profession is tight enough that nearly three-quarters of the members of the latest *DSM* task force had ties to drug companies.[25] Moreover, pervasive drug advertisements are probably the most significant conduit of information to the general public about *DSM* diagnoses.

Social factors have completely transformed the value of diagnoses over the *DSM* period. When the first manual arose in 1952, diagnoses were typically applied to involuntarily confined patients. By the mid-1960s, however, the population of mental hospitals was steeply declining at the same time as the number of people seeking voluntary forms of outpatient treatment was dramatically escalating. The anti-psychiatry narrative that gained credence in the 1960s and early 1970s was becoming outdated. By the 1980s coercive labeling was rare, and diagnoses were more likely to be valued than opposed. Patients and their families sought them to obtain desired treatments, reimbursement for care, eligibility for government benefits, provision of special education resources, and explanations for distress. One psychiatrist notes in regard to ADHD, "And the parents are pushing you to do something, and they even bring up the subject of medication by themselves, and they know all the criteria for ADHD, and they've read books and have been all over the internet on this thing."[26] In other words, diagnoses are intensely *reactive* to consumer needs. "There is an enormous pressure on psychiatrists and other physicians to diagnose," David Kupfer, chair of the *DSM-5* revision observes.[27]

These internal and external pressures force the developers of the various *DSMs* to be highly sensitive to collective social forces. Professional organizations, advocacy groups, clinicians, and researchers exert political pressure that pushes the manual to include or exclude various diagnoses. Economic concerns drive mental health professionals, pharmaceutical companies, and consumers and their families to broaden the scope of classificatory systems. Health care administrators and managers require diagnoses to rationalize their provision of services. All these factors have undergone fundamental changes over the course of the *DSM* era. The changing nature of the *DSM* inevitably reflects professional, political, cultural, and economic forces.

The Diversity of Diagnoses

There is no single history for the major diagnostic classes in the *DSM*. Instead, intra-professional and external interests have varied influences on different diagnoses. Eight of the most prominent *DSM* classes are the psychoses, anxiety and depressive neuroses, posttraumatic stress disorder, personality disorders, substance use disorders, gender- and sexuality-related behaviors, and a variety of conditions found among children and adolescents. Social forces have shaped the *DSM*'s portrayal of these conditions in diverse ways.

The major psychoses—schizophrenia, bipolar disorder, and melancholia—are the most easily recognizable psychiatric disorders. They are marked by behaviors, thoughts, and emotions that seem disconnected from reality and are often accompanied by delusions and hallucinations. Ancient medical depictions of these conditions provide recognizable variants of our current understanding of these illnesses.[28] Despite numerous modifications to the details, descriptions of the psychoses have remained fairly constant across the *DSM* era. As psychiatric practice shifted from mental hospitals to outpatient treatment, however, the central-ity of the psychoses in the *DSM* has correspondingly declined.

In contrast to the psychoses, characterizations of the various neuroses have dra-matically changed over the course of diagnostic history. This class includes dis-tressing conditions such as depression, anxiety, and psychosomatic ailments. Since the seventeenth century, people with complaints much like what we now call de-pression and anxiety have commonly sought medical help. The first widely known condition of this nature was the "English malady," a term coined by English physi-cian George Cheyne in 1733 to refer to a mélange of psychosomatic, nervous, and depressive symptoms. In the second half of the nineteenth century, American neurologist George Beard named this potpourri "neurasthenia," and it became an extremely popular diagnosis in both the United States and Europe. The neuras-thenic package of fatigue, sadness, anxiousness, and other stress-related com-plaints underlies many *DSM* diagnoses.[29]

PTSD is a relatively new type of mental disorder, first diagnosed in the latter decades of the nineteenth century.[30] It is unique among current diagnostic classes because the name itself definitionally indicates its cause. It is impossible for some-one who has not experienced some environmental trauma to receive the diagnosis. Therefore, criteria must both specify the nature of the traumas that give rise to PTSD and define its symptoms. Because stressor-related conditions are tied to an external precipitant, they are among the least stigmatized psychiatric diagnoses.

The personality disorders, too, are a relatively new class. Unlike other catego-ries, they do not feature particular symptom clusters but instead reflect enduring character traits that define what kind of person someone is. They first attracted widespread psychiatric attention during the 1930s in the writings of analytic fol-lowers of Freud.[31] This class is more important to clinicians than to researchers, who often scorn it as being difficult to measure and to distinguish from normal personality styles. Intra-professional interests have dominated *DSM* portrayals of the personality disorders because no drugs are specifically designed for them and no advocacy groups lobby for their inclusion in the manual.

Psychoactive substances have a long and complicated relationship to psychiat-ric practice. On the one hand, at least since the time of the Hippocratic physicians in ancient Greece, drugs such as opium and alcohol have been widely employed to treat almost all mental disorders. On the other hand, the same drugs that some-times effectively treat one disorder can produce another. In 1804 Scottish physician Thomas Trotter proposed that the underlying pathological quality of substance use was "the evil genius of the habit."[32] The qualities Trotter labeled as "habit" for when substance use becomes compulsive and uncontrollable are now commonly called "addiction" or "dependence." As much as any other major category of mental disorder, diagnoses of substance use and abuse have long been intertwined with moral and legal considerations.

Cultural values also fundamentally shape diagnoses related to gender and sexuality. Their criteria are intrinsically connected to evaluations of what are "normal" or "abnormal" sexuality and gender identity. Diagnostic portrayals of this class have substantially changed as norms regarding gender and sex drastically altered over the DSM era. Unsurprisingly, these changes have produced some of the most divisive controversies in the manual's history.

Mental disorders among youth have grown more than any other class over the DSM period. Before 1980 the manual mostly limited childhood disorders to what it called "mental retardation." Especially since the DSM-IV in 1994, this class of disorders expanded more rapidly than any other as epidemics of ADHD, autism, and childhood bipolar disorder arose. This class is unique because young people themselves rarely initiate help-seeking efforts. Instead, parents and, occasionally, other adults seek treatment for disturbed youths. Families highly value DSM diagnoses for their children because they bring about desired treatments and educational services. Drug companies, too, have developed intense interest in a class of mental disorders that arises early in life, persists for decades, and seems to require daily doses of medication. Because immaturity, temper tantrums, and oppositional behaviors are common in childhood and adolescence, mental disorders among the young are often difficult to distinguish from normal mood difficulties that typically arise at this age.

The distinct trajectories of different conditions across the DSM era means that the manual does not have one common history. The value placed on diagnoses, the types of intra-professional concerns, and the degree of involvement of external interests vary considerably across the main classes of psychiatric illness. Therefore, no single history captures the various paths of the mixture of phenomena that the DSM classifies as "mental disorders."

An Overview of the History of the DSM

This book traces the evolution of the DSM from its origin in 1952 through its latest edition, the DSM-5, in 2013. It groups the history of the manual into three major stages. The following chapter discusses how the first two DSMs adeptly integrated the organic and psychotic conditions found among institutionalized patients with the neuroses and personality disorders that dominated surging outpatient practices. No one paid them much attention, however, because diagnoses had few uses either inside or outside of the psychiatric profession.

The next three chapters analyze the *DSM*'s second stage, which emerged with the *DSM-III* revolution in 1980 and persisted through the *DSM-III-R* (1987) and *DSM-IV* (1994) revisions.[33] Intra-professional needs for psychiatry to be recognized as a legitimate branch of medicine, far more than any new knowledge, drove this diagnostic transformation. External pressures to conform to the demands of the insurance industry, government agencies, advocacy groups, and drug companies also contributed to the conversion of what had been cursory descriptions into well-defined, highly specific diagnoses.

Chapter 6 covers the third and, to date, final stage of the manual's history. It arose when researchers realized that the *DSM* entities hindered rather than helped their efforts to understand the nature of mental disorders. The leaders of the revision did not, however, grasp the importance that the manual had assumed for clinical and administrative practice. Reversing the dynamics of the *DSM-III* process, clinicians were so invested in *DSM* diagnoses that they beat back a researcher-driven attempt to make fundamental changes in the classification. The very public disputes that accompanied the *DSM-5* revision shattered the credibility of the manual but failed to produce any viable alternative to it.

The final chapter sums up this history's major implications for psychiatric diagnoses. Over the course of the *DSM* period, the manual changed from a minor and rarely used volume to an authoritative classification with huge influence both inside and outside of psychiatry. More than a steady progression of scientific knowledge or a growing amount of mental pathology, the various editions of the *DSM* reflect shifts in intra-professional dynamics, economic incentives, and political power. These changes show not only how the *DSM* is a product of broader collective forces but also how the manual itself has shaped perceptions of mental disorder. Definitions of mental disorders, like all classifications, are both products of and influences on the social world in which they are embedded.

TWO | The *DSM-I* and *DSM-II*

Classifications of psychiatric diagnoses underwent huge changes over the course of the nineteenth and twentieth centuries. Before the 1950s they reflected the centrality of inpatient hospitals in the mental health system. As psychiatric practice moved out of asylums and into communities, the first *DSM*s arose to accommodate a new type of outpatient treatment. Once psychiatry's legitimacy came to depend on its identification with scientific medicine, the *DSM-I* and *II* diagnostic system became anachronistic and ripe for a complete reorientation.

Before the *DSM*

Attempts to categorize disturbed mental states have ancient roots. For thousands of years, mental suffering was viewed as reflecting an imbalance of forces within individuals and between individuals and their environments. The dominant Hippocratic/Galenic medical tradition postulated that four basic temperaments corresponded to the dominant fluids (what it called "humors") in bodies. It associated enthusiastic and easily excited sanguine personalities with high levels of blood; confident and determined choleric types with dominance of yellow bile; quiet and anxious temperaments with marked amounts of black bile; and, finally, thoughtfulness and even temper with high quantities of phlegm. Excessive

or deficient amounts of each humor led to undesirable mental traits such as im-
pulsivity, aggression, melancholy, or apathy. Hippocratic conceptions of disease as
disturbing the balance between nature, society, and the individual persisted
among general physicians and the lay public into the nineteenth century.[1]

In contrast to the highly generalized Hippocratic notions, concepts of disease
specificity are relatively new. With but a few exceptions such as smallpox or yellow
fever, premodern medicine rarely discerned distinct categories. For mental disorders,
too, a small number of capacious conditions such as melancholia or mania sufficed as
classifications. The first American psychiatrist, Benjamin Rush, went even further:
for Rush almost all ailments could be reduced to manifestations of the single disorder
of fever.[2] Likewise, treatments for psychic ills were not illness specific but applied
across a wide range of conditions. Physicians strove to heal the whole person rather
than provide some distinct remedy for a specific problem. Therefore, early classifications of mental disorders had little importance or influence.

In Europe and the United States around the turn of the eighteenth century,
psychiatry arose as a freestanding discipline dedicated to the identification, care,
and treatment of mental illness (then usually called "madness" or "insanity"). The
first psychiatrists (then called "alienists" or "mad doctors") were associated with
newly developed institutions that housed people with serious mental disorders.
Their pragmatic needs to classify patients upon admission, assign them to particular wards, and release them with some label led to the emergence of more explicit diagnoses. Bureaucratic necessities—as opposed to understanding the nature
of various mental disorders—therefore dominated initial categorizations.

ASYLUM-BASED CLASSIFICATIONS

Most early psychiatric diagnostic schemes contained only a small number of
categories.[3] Because the causes of mental illness were unknown, the initial groupings relied on observable symptoms to differentiate diverse conditions. "When
the science of causes shall be complete we may then make them the basis of our
classification, but till then we ought to content ourselves with an arrangement according to symptoms," British alienist Thomas Arnold wrote in 1782.[4] French
psychiatrist Philippe Pinel produced the first recognizable predecessor to modern
classification systems. It contained just four entities: mania (conditions marked by
excitement and fury), melancholia (serious depression), dementia (incoherent
thoughts), and idiocy (intellectual deficiency and organic dementia).[5] At the time,
psychiatric labels resembled medical categories, which were equally broad based
before the emergence of modern bacteriology in the late nineteenth century.

Nineteenth-century American alienists were for the most part uninterested in elaborate classification systems.[6] In their eyes, detailed groupings were of little utility and did not adequately portray the protean symptoms of insanity. Moreover, they believed that effective therapies had to reflect each person's unique circumstances rather than any general principle. Psychiatrists occasionally debated the validity of such specific categories as moral insanity (a condition in which there was a morbid perversion of the emotions but little or no impairment of the intellect), but this was the exception rather than the rule.

Early psychiatrists and neurologists relied on the only resource they possessed—external signs and symptoms—to make diagnoses of mental disorder. The problem they faced was that mental illness has so many symptoms in such variety that innumerable configurations were possible. For this reason, a major figure in American psychiatry during the last half of the nineteenth century, Pliny Earle, claimed that defining insanity was an "impossibility."[7] The intellectual and scientific constraints on the development of psychiatric categorizations persisted for much of the century. When Earle was queried in 1886 about the possibility of developing a universally accepted classification of mental diseases, he replied in negative terms: "In the present state of our knowledge, no classification of insanity can be erected on a pathological basis . . . [since] the pathology of the disease is unknown. . . . We are forced to fall back on the symptoms of the disease—*the apparent mental condition*, as judged from the outward manifestations."[8] Pinel's traditional groups of mania, melancholia, dementia, and idiocy, therefore, still sufficed.

Other alienists believed that even Pinel's four categories were different expressions of the unitary psychosis of "insanity."[9] The most influential American psychiatrist of that era, Isaac Ray, denied that any classification system could be "rigorously correct, for such divisions have not been made by nature and cannot be observed in practice."[10] At best, Ray wrote, insanity could be divided into two broad groups. The first—idiocy and imbecility—was composed of individuals with congenital defects. The second encompassed those with lesions that impaired the functioning of their mind, including mania and dementia. The US Census of 1880 was somewhat more specific, using seven types of mental disorders: mania, melancholia, monomania, paresis, dementia, dipsomania, and epilepsy.[11] Psychiatrists made little diagnostic progress over the course of the century. In 1892 prominent asylum keeper Hack Tuke summarized, "The wit of man has rarely been more exercised than in the attempt to classify the morbid phenomena covered by the term 'insanity.' "[12]

Outside of mental institutions, the general physicians and neurologists who treated patients with less severe ills were equally uninterested in specific diagnoses. In 1869 American neurologist George Beard developed what became the overwhelmingly popular concept of "neurasthenia." This commodious term, which remained in widespread use in both the United States and Europe through World War I, captured an extremely broad range of symptoms, including general malaise, fatigue, tension, anxiety, depression, and low mood. It encompassed what would later be seen as anxiety neuroses, depressive conditions, and a host of psychosomatic ills. In line with then-current medical theory, Beard posited that neurasthenia reflected an underlying weakness of nervous energy and so had a somatic basis.[13]

Toward the end of the nineteenth and beginning of the twentieth centuries, interest in psychiatric classification intensified as clinicians began to shift their attention away from symptomatic presentations to the course and outcome of mental disorders. A major stimulus for this new diagnostic emphasis stemmed from the remarkable progress in determining the cause, prognosis, and treatment of syphilis (originally called "general paralysis of the insane").[14] This widely prevalent disease accounted for about a fifth of all admissions to mental hospitals in the nineteenth century. Its symptoms, including delusions of grandeur, hallucinations, confused thinking, and periods of deep melancholy, were usually indistinguishable from psychotic and other mental disorders. In a remarkable series of developments over a six-year period beginning in 1905, the bacterium *Treponema pallidum* was found to cause syphilis, a blood test to identify this condition was invented, and an effective treatment for it was discovered.[15] The example of syphilis showed how a condition that appeared to be a mental disorder was actually a genuine medical disease. It gave hope that many other psychiatric conditions might also have organic causes, be identified at an early stage through an objective test, and be prevented from progressing in severity.

The second breakthrough in psychiatric classification stemmed from the asylum-based work of German psychiatrist Emil Kraepelin. Kraepelin's major accomplishment was to turn attention from the visible symptoms of mental disorders to their course and outcome. Using thousands of detailed case histories, Kraepelin deemphasized the symptoms that are present at a particular point in time and made prognosis—the natural history of a disorder—the central organizing principle of his diagnostic system.[16] He published the first edition of his famous text, *Compendium of Psychiatry: For the Use of Students and Physicians*, in 1883. In its fifth edition (1896), Kraepelin spoke of a "decisive step from a symptomatic to a clinical view of insanity. . . . The importance of external clinical signs has . . .

been subordinated to consideration of the *conditions of origin*, the *course*, and the *terminus* which result from individual disorders. Thus, all purely symptomatic categories have disappeared from the nosology."[17]

Kraepelin emphasized only a small number of basic psychiatric diagnoses. In particular, he singled out the specific disease of dementia praecox (what is now called schizophrenia)—a very severe psychosis that generally arose among young adults and caused progressive deterioration—and its three subtypes of paranoid, catatonic, and hebephrenic.[18] In the text's sixth edition (1899), he distinguished manic-depressive psychosis as a distinct condition marked by mood swings that had a varying rather than worsening course. This disorder "includes on the one hand the entire territory of the so-called periodic and circular insanities, on the other hand, simple mania that hitherto has been kept separate. Over the years I have become increasingly convinced that all these pictures are only presentations of a single disease process."[19] In his later works Kraepelin came to consider any distinctions between manic depression and unipolar depression inessential because both were different manifestations of the same underlying disease. He combined all severe depressive disorders, whether or not they cycled with mania, into a single diagnosis.[20] Around the same time Swiss psychiatrist Eugen Bleuler substituted the term "schizophrenia" for "dementia praecox" because he believed that a splitting of psychic functions, rather than dementia, constituted the essence of this condition.[21]

Kraepelin's lasting diagnostic influence was to emphasize what different patients had in common, thus diverting attention from unique circumstances toward disease processes. Both dementia praecox and manic depression were natural entities that arose independently of the particular experiences of patients or the diagnostic preferences of clinicians. Kraepelin viewed many as inherited brain diseases that worsened as they were transmitted across generations and others as results of external toxins.

In general, early twentieth-century American psychiatrists displayed less interest in classification than did their European counterparts. The Kraepelinian emphasis on specific diseases was not immediately accepted in the United States, where diagnostic uncertainty persisted. The only widely accepted distinction was between the two very general categories of psychosis and neurosis.[22] The authors of two major American textbooks, Henry J. Berkley and Stewart Paton, conceded that classification in psychiatry and medicine differed; the former was unable to create an etiologically based system because the indications of one form of mental disease overlapped so highly with others.[23] In practice, each separate mental institution relied on its own classification of diseases, resulting in a "polyglot of diagnostic labels and systems, effectively blocking communication and the collection of medical statistics."[24] Indeed, the proliferation of classifications led Charles G. Hill to observe in 1907 that there was little room for additional diagnoses "unless we add 'the classifying mania of medical authors.'"[25]

At the same time as Kraepelin was pursuing his intense interest in classifying specific types of mental disorders, Sigmund Freud and the immensely popular school of psychoanalysis he founded were moving in the opposite direction. Although Freud's early work in the 1890s, like Kraepelin's, was devoted to distinguishing different forms of neuroses, around the turn of the century his thought turned sharply away from classificatory issues.[26] Instead, he came to focus on the unconscious psychic conflicts that presumably lay behind the array of overt manifestations. Indeed, for Freud, examining external signs diverted attention from the far more essential issue of determining the symbolic functions of symptoms in expressing unresolved childhood issues. The rise of psychoanalysis to dominance in American psychiatry in the middle decades of the twentieth century meant that outpatient practitioners had little concern with diagnostic issues.

The rise of the mental hygiene movement in the 1920s and 1930s also ensured that diagnoses would not hold much interest for American psychiatrists who practiced outside of mental institutions.[27] Mental hygienists established outpatient clinics, child guidance practices, and treatment facilities in schools and colleges. They focused on promoting positive mental health, preventing mental illness, and alleviating stressful conditions in the social environment. None of these activities required specific diagnoses, so mental hygienists, like psychoanalysts, had little motivation to develop classifications of mental disorders. At the same time, psychotropic drugs that were used for common mental problems, such as the barbiturates and stimulants, had general indications unrelated to particular diagnoses.[28]

The Statistical Manual

Throughout the first half of the twentieth century, the great majority of American psychiatrists worked in institutions, employing organic responses, including sedative drugs, insulin coma injections, shock treatments, fever therapies, and surgical procedures, such as lobotomies, across a wide range of conditions.[29] At best, diagnoses were of marginal significance for asylum treatments. Mental hospitals, however, were the largest items in most state budgets and had to account to legislatures for how they spent their funds. These institutions had to sort incoming patients into appropriate groups, assign them to suitable wards, and categorize them when they left. This involved producing records of the number and types of patients they admitted, treated, and discharged. Yet each hospital had its own method of accounting, and no uniformity existed across hospitals or different states.[30]

The initial American diagnostic manual emerged from the highly practical management needs of asylums. In 1918 the American Medico-Psychological Association (subsequently renamed the American Psychiatric Association in 1921) issued the first standardized psychiatric guidebook, the *Statistical Manual for the Use of Institutions for the Insane*. "At the time that this classification was adopted . . . there were about as many statistical outlines in use in this country as there were mental hospitals," psychiatrist Karl Menninger later observed.[31] The manual reflected the conditions found in institutionalized populations as well as the belief that mental disorders had a biological foundation. It did not, however, employ Kraepelin's procedure of separating conditions through their differing courses or prognoses. The *Statistical Manual* guided psychiatric classification in the United States through the end of World War II.

The *Statistical Manual* contained 21 principal groups (and an additional group of "not insane"). Twenty of these were psychoses (e.g., psychoses with brain tumor, alcoholic psychoses, dementia praecox), and just one—psychoneuroses—was non-psychotic.[32] The neurotic category mentioned just four subtypes: hysterical, psychasthenic (obsessive compulsive), neurasthenic (fatigue), and anxiety. None of its diagnoses was based on empirical data; each was a category that seemed to reflect a consensus among asylum psychiatrists that their descriptive character was adequate (e.g., "psychoses due to unknown or hereditary causes, but associated with organic change").

Psychiatrists were only marginally concerned with the *Statistical Manual*'s diagnostic system; its purpose was to facilitate the collection of statistical data in mental hospitals. This first uniform psychiatric classification attracted much criticism. For example, Adolf Meyer, the most prominent American psychiatrist during the first half of the twentieth century, thought that the *Statistical Manual*'s diagnoses were worthless.[33] Meyer was devoted to studying individual lives that he believed could not be reduced to disease types. Although he developed his own sixfold classification of reaction types, Meyer's psychobiology, which attempted to integrate individual life experiences with psychological and biological data, did not lend itself to rigid diagnostic categories. Nevertheless, the *Statistical Manual* became the definitive nosology of the interwar years and went through 10 editions between 1918 and 1942. Each edition continued to emphasize conditions found among institutionalized patients as well as maintain a somatic viewpoint.

MEDICAL 203

World War II led to a major turning point in psychiatric diagnosis. In the initial mobilization for the war, psychiatrists were employed to screen millions of volunteers and draftees who were neither institutionalized nor seeking outpatient therapy for signs of mental illness. They rejected nearly 2 million draftees diagnosed with presumed disorders. The war itself exposed many of these seemingly normal soldiers to extreme environmental stressors. As a consequence, huge numbers suffered mental breakdowns. Yet most of these casualties fell outside of the extant *Statistical Manual*'s range of disorders. Only a small fraction featured the psychotic conditions that were at the center of prewar interest.[34] To describe their conditions, psychiatrists resurrected "shell shock" from World War I or used nonspecific terms such as "combat fatigue" or "war neurosis."

Yet the bureaucratic nature of military institutions required some means of classifying the reasons for all types of casualties:

> The Armed Forces faced an increasing psychiatric case load as mobilization and the war went on. There was need to account accurately for all causes of morbidity, hence the need for a suitable diagnosis for every case seen by the psychiatrist, a situation not faced in civilian life. Only about 10% of the total cases seen fell into any of the categories ordinarily seen in public mental hospitals. Military psychiatrists, induction station psychiatrists, and Veterans Administration psychiatrists, found themselves operating within the limits of a nomenclature specifically not designed for 90% of the cases handled.[35]

This situation shone a bright light on the limitations of the institution-based *Statistical Manual* and made apparent the need for a new psychiatric classification system.

William Menninger (Karl's brother) was "the acknowledged national spokesperson of American psychiatry until his death in 1966."[36] During the war, Menninger rose to brigadier general, the highest military rank any psychiatrist had ever held. In 1944 he developed a set of standards for military physicians who treated psychiatric casualties. His recommendations were almost purely psychoanalytic, emphasizing unconscious conflicts, infancy and childhood molding adult traits, and insight in the therapeutic relationship. He then led an interdisciplinary group of mental health professionals, including clinical psychologists, social workers, and psychiatrists, to develop a new classification system that better fit the problems that wartime psychiatric casualties displayed.[37]

The manual this group produced, *Medical 203*, was published in 1946 with the intention of bringing it to widespread civilian as well as military use.[38] In thoroughgoing contrast to the *Statistical Manual*, psychosocial principles that emphasized the interrelationship between an individual's life history and social environment framed *Medical 203*. The first of its five categories, transient personality reactions to acute or special stress, displayed priorities that were the diametric opposite of the *Statistical Manual's* concern with hospitalized patients. Originating from the types of psychic casualties in the military, it applied to previously normal people who developed brief neurotic symptoms in the face of overwhelming stress. These reactions were usually transient and reversible. Each of these assumptions—that the afflicted were ordinary people, their problems stemmed from some external precipitant, and their conditions were not chronic—starkly contrasted with the diagnoses found in the extant manual. Yet this category was at the core of *Medical 203*.

The second category, psychoneurotic disorders, also exemplified the differences of this classification from the *Statistical Manual's* conditions. The psychoneuroses were explicitly grounded in Freudian psychodynamics, resulting from repressed emotions in infancy and childhood. The chief characteristic of each of the eight neuroses was anxiety, which was expressed through various defense mechanisms. Following Meyer, these subcategories—anxiety, dissociative, phobic, conversion, somatization, obsessive compulsive, hypochondriacal, and depressive—were all called "reactions" because they represented different types of responses to developmental experiences.

The third class, character and behavior disorders, comprised pathological personality types. Unlike the disorders in the first two categories, they did not entail distress but used an individual's conduct as the basis of classification. Such disturbances rarely came to the attention of hospital-based psychiatrists in civilian settings, but they were potentially important sources of disruption within closely knit combat units. Yet the only label from the *Statistical Manual* that could have been applied to them was "psychoses with psychopathic personality." *Military 203* adopted a much broader conception of such conditions: its three subtypes encompassed seven kinds of pathological personality types (schizoid, paranoid, inadequate, antisocial, asocial, sexually deviant, and cyclothymic emotional swings), addictions, and five kinds of immaturity reactions. All referred to characterological problems that were absent from the prewar manual.

The fourth class, disorders of intelligence, was the only category that often had an organic etiology. Brain diseases caused some of these cases, while other mental deficiencies had unknown etiologies. A "failure to test and evaluate correctly external reality" characterized the final class, psychotic disorders. In Kraepelinian fashion, it grouped patients into three general categories of schizophrenic, paranoiac, and affective psychoses. Departing from Kraepelin, however, the manual assumed that, aside from a residual category of psychoses with demonstrable changes in brain structure, these types had no known organic etiology. Indeed, the manual's only reference to Kraepelin states that "it is not essential forcibly to classify such patients into a Kraepelinian type."[39]

Medical 203 was also notable for anticipating the multiaxial system that would later reemerge in the *DSM-III* (1980). It insisted that a complete diagnostic evaluation involved four considerations: the type and severity of symptoms, the amount of stress, the presence of personality types that predisposed someone to develop a mental illness, and the degree of incapacity. This manual also initiated a type of dimensional measurement: physicians were to record each diagnosis as not present or mild, moderate, or severe. It did not, however, discuss the relationship between a diagnosis and any particular treatment.

Medical 203 became widely used soon after it was published. However, competing classification systems, including the *Statistical Manual*, various military taxonomies, a listing of psychiatric disorders in the American Medical Association's *Standard Classified Nomenclature of Disease*, and the *International Statistical Classification*, created an untenable and confusing diagnostic picture.[40] The US psychiatric profession clearly needed some standardized classification system that met the needs of all of its members.

From Hospital to Community

Psychiatric practice and training were undergoing dramatic changes at the time *Medical 203* became available. Serious internal divisions marked the field during the immediate postwar period. When World War II began, about two-thirds of psychiatrists worked in mental institutions where they used biological accounts of and somatic treatments for mental illnesses. Hospital-based psychiatrists held a narrow agenda that focused on the institutions treating the severely mentally ill, featured an organic model, and avoided broader social issues. A younger group of psychiatrists, who had served in World War II, regarded the asylum-oriented practitioners as outmoded relics of a bygone era. The new generation was more in line with the earlier mental hygiene movement: these psychiatrists assumed that mental illness had psychosocial causes, favored a broad psychodynamic model of mental health, and advanced sweeping proposals for social change.

In the postwar era, psychiatry shed its traditional preoccupation with persons with severe and persistent mental illnesses and became most concerned with the psychological difficulties of a far larger and more diverse outpatient population as well as with social problems more generally. By 1956 only about 17 percent of the 10,000 or so APA members were employed in inpatient settings.[41] The shrinking group that practiced in mental hospitals had little prestige and was considered a remnant of an anachronistic biological tradition.[42] Psychiatrists increasingly shifted their focus away from the psychoses and toward the less severe and more common neurotic conditions found among outpatients. A study of 100 psychiatric patients in three private practices that was published in a 1950 issue of the *American Journal of Psychiatry* indicated the trend in outpatient labeling at the time. Just over half received a diagnosis of some psychoneurosis, 26 of a psychosis, and 22 did not receive a diagnosis.[43]

At the same time, psychiatric theories were undergoing radical changes. The Nazi rise to power in Germany and Austria propelled an emigration of Freudian analysts to the United States, where the dominance of Meyer's eclectic psychobiology provided an accommodating environment for them. His model downplayed particular diagnoses in favor of understanding all forms of mental illness as types of individuals' maladjustments to their circumstances. Meyer was far more interested in individual "reactions" to their biology, psychology, and social environments than in determining what particular diagnosis they might qualify for.[44] "My first advice was to let classification rest or to give it a secondary place," he wrote.[45] His perspective—as opposed to Kraepelin's focus on the future course of symptoms—led clinicians to look backward to the life histories that led patients to become ill. Moreover, in contrast to asylum psychiatrists, Meyer did not assume that mental health and disorder were distinct but instead viewed them as points on a continuum. Not just knowledge about but also treatments for problems depended on the particular life experiences of each patient. This orientation placed psychiatry in fundamental opposition to other branches of medicine, which were increasingly dominated by notions of disease specificity.

Like Meyer's psychobiology, psychoanalysis eschewed diagnosis. Prominent psychiatrist Roy Grinker noted the "aversion to diagnosis" among analysts.[46] A leading analyst Franz Alexander observed, "Each case is a unique problem. What the therapist is primarily interested in is not the nosological classification of a person, not in what way he is similar to others but in what way he differs from them."[47]

Although psychoanalysts were few in number, they had a disproportionate influence on American psychiatry and culture. In 1946, William Menninger led a group of psychiatrists who had served in World War II in forming the Group for the Advancement of Psychiatry (GAP), which strove to use psychodynamic principles as the basis for advocating widespread social change. One of GAP's founding documents stated that the group would apply "psychiatric principles to all those problems which have to do with family welfare, child rearing, child and adult education, social and economic factors which influence the community status of individuals and families, inter-group tensions, civil rights and personal liberty. This, in a true sense, carries psychiatry out of the hospitals and clinics and into the community."[48] GAP's activist agenda could hardly have been a more resounding contrast with the institutional and organic thrust of asylum psychiatry. Within two years, psychiatrists affiliated with GAP had gained control of the APA.[49]

Psychiatric education, too, dramatically changed in the postwar period. Anticipating the *DSM-I*, an APA conference in 1951 urged that medical students should be trained "to diagnose correctly the condition of patients who are emotionally disturbed and who may be expressing their distress in physical, psychological or social symptoms. This ability implies . . . a reasonable understanding of the zones of healthy and sick behavior in our society and, more particularly, the ability to differentiate between normal, neurotic, psychopathic, psychotic, and intellectually defective behavior."[50] These broad distinctions were sufficient diagnostic resources to classify the huge range of problems that led people to enter outpatient therapy.

By the early 1950s, psychodynamic psychiatrists had prevailed over biologically oriented ones as analysts gained control of psychiatric education. Psychoanalysts, who had been almost totally absent from the academy in the prewar period, came to dominate departments of psychiatry in medical schools.[51] Correspondingly, few new psychiatrists took jobs in mental hospitals: a third worked in psychiatric clinics, nearly a third joined academic institutions, and most of the final third entered private practice. In addition, dynamically oriented psychiatrists dominated the process leading to the first diagnostic manual of the APA.[52]

The *DSM-I*

Retrospectively, the *DSM-I* seems to have been a cursory and perfunctory classification of mental disorders.[53] Yet, at the time, the manual was a marked advance in psychiatric diagnosis. The *DSM-I* helped to reconcile the competing diagnostic forces in play during the postwar period. Perhaps its major accomplishment was to encompass both the *Statistical Manual*'s focus on conditions typically seen in mental institutions and *Medical 203*'s emphasis on the psychosocial problems found among outpatients.

Hospital-based psychiatrists required a system that would meet their need to classify inpatients. As of June 30, 1950, there were 577,000 patients in all hospitals for the prolonged care of the mentally ill in the United States. This number had steadily increased every year over the previous decade and was nearing the all-time high it would reach in 1954.[54] There was no way for anyone to know that the population of mental hospitals would soon begin a steep decline.

The dramatic changes in psychiatric theory and practice that followed World War II also demanded a new diagnostic system. The profession's traditional attention to people with severe mental disorders was giving way to a concern with the psychological problems of a far larger and more diverse group. Psychodynamic psychiatrists—who had neither need nor esteem for diagnoses—were reshaping the specialty and assuming a dominance that would persist for two decades. They transformed the role of psychiatrists from managing inpatient institutions to promoting widespread changes that might improve mental health in the general population. Psychiatrists increasingly shifted their activities away from the psychoses toward milder conditions in the hope that early treatment of functional but distressed individuals would ultimately diminish the incidence of the more serious disorders.[55]

Psychiatrists who adhered to Meyer's psychobiology emphasized the role of factors such as difficulties in parent-child relationships and marriage, emotional immaturity, and the inability of individuals to adjust to their environments in bringing about mental disturbance. Such concerns with life histories did not lend themselves to specific diagnostic categories. Psychodynamic psychiatrists, too, minimized diagnostic categories and emphasized the need to focus on a broad range of conditions. Psychoanalysts, a growing force in the profession, strove to understand the underlying psychic conflicts that produced a diffuse array of symptoms that varied across individual patients. By its very nature, analysis was incompatible with specific diagnoses. Karl Menninger proclaimed, "There is only one class of mental illness—namely mental illness."[56]

Under these circumstances, in 1948 the APA began to consider developing a new classification system that incorporated psychodynamic concepts. This effort also involved a number of governmental agencies, including the US Army, the Veterans Administration, and, from 1950, the National Institute of Mental Health. Their efforts resulted in the publication of the first *Diagnostic and Statistical Manual (DSM-I)* in 1952. The new manual reflected the intellectual, cultural, and social forces that had transformed psychiatry during and after the war. It accommodated the needs of both hospital and outpatient psychiatrists while making it clear that psychiatry's center of gravity had moved from institutional to community practice. The *DSM-I*'s "basic division" was between mental disorders that resulted from impairments of brain function and those that stemmed from difficulties in individual adaptation to the environment.[57] In this way, the *DSM-I* deftly combined the *Statistical Manual*'s focus on organically grounded psychoses with *Medical 203*'s emphasis on stress-related, neurotic, and personality conditions.

The first group of *DSM-I* disorders consisted of cases in which some damage to brain tissue produced or precipitated the disturbed mental function. All of the 26 listed syndromes entailed impairments of orientation, memory, intellectual functions, and judgment as well as unstable moods.[58] These conditions were split into 11 types of acute, potentially reversible disturbances and 15 types of syndromes involving permanent brain damage. Each type was classified by its cause, usually an organic disease, poison, or drug or alcohol addiction (e.g., acute brain syndrome associated with intracranial infection, chronic brain syndrome associated with central nervous system syphilis) along with a residual category of unknown cause. The organic category also contained two diagnoses of acute and chronic brain disorders caused by external trauma. All the syndromes could be mild, moderate, or severe.[59] Despite the 26 distinct diagnoses, organic conditions were far more limited in scope than they had been in the *Statistical Manual*.

Medical 203 was the major immediate source of the other basic *DSM-I* delineation: psychogenic disorder.[60] The new manual left no doubt that explanations assuming a condition had a psychological, as opposed to a physical, cause had replaced the organic thrust of the previous *Statistical Manual*. Reflecting the experiences of military psychiatrists who had dealt with millions of traumatized soldiers in World War II, the psychogenic category used a general framework that blended a Meyerian eclecticism, focusing on the interactions between vulnerable personalities and stressful life events, with Freudian psychodynamics, emphasizing unresolved, unconscious conflicts that arise in early childhood. In line with then-current psychiatric thinking, the *DSM-I* assumed that psychogenic disorders resulted when environmental stressors evoked individual predispositions developed in accord with analytic principles.[61]

The manual divided psychogenic disorders into five major types: psychotic, psychophysiologic, psychoneurotic, personality, and transient situational personality disorders. In contrast to the organic category, these classes had little specificity, stemming from more general inabilities of individuals to adjust to their life circumstances. Therefore, the manual used the term "reaction" to characterize all of these types of disorders.

The *DSM-I*'s placement of the core psychotic conditions in the psychogenic category indicates the dominance of psychodynamic approaches in the postwar era. Although the manual makes no reference to dementia praecox, instead using Bleuler's term, "schizophrenia," it reflected the tripartite Kraepelinian division of schizophrenia, manic depression, and paranoia.[62] In addition, its division of schizophrenia into nine subtypes—including Kraepelin's hebephrenic, catatonic, and paranoid versions—echoed his emphasis on distinguishing specific conditions from each other. Yet their psychogenic aspect contrasted with Kraepelin's insistence that psychoses were grounded in organic disturbances. In addition, it stated that the basic pathology of the psychoses involved "struggle for adjustment to internal and external stresses," a dynamic far removed from Kraepelin's biological conceptions but in accord with Meyerian and analytic views.

Psychosomatic medicine had exploded in popularity after the Second World War and remained influential through the mid-1960s.[63] The placement of psychophysiologic disorder as the manual's second psychogenic category reflects this status. At the time, psychosomatic assumptions were applied to a broad range of physical problems whose cause could not be determined—ulcers, asthma, hypertension, arthritis, and many others—but that were assumed to arise from a conversion of unconscious conflicts into somatic forms. This category was explicitly framed through dynamic principles: "These reactions represent the visceral expression of affect which may be thereby largely prevented from being conscious. The symptoms are due to a chronic and exaggerated state of the normal physiological expression of emotion, with the feeling, or subjective part, repressed."[64] Each of the 10 specific psychosomatic reactions was associated with a different bodily organ—for example, respiratory, cardiovascular, or gastrointestinal reactions.

The third class of psychogenic disorder, the seven psychoneurotic reactions, was also unambiguously formulated through psychodynamic principles. In contrast to Kraepelin, who elevated depression and considered anxiety marginal to psychiatric diagnosis, the *DSM-I* followed Freud in placing anxiety at the heart of all the neuroses. In all seven diagnoses "anxiety is a chief characteristic, directly felt and expressed, or automatically controlled by such defenses as depression, conversion, dissociation, displacement, phobia formation, or repetitive thoughts and acts."[65] For example, the *DSM-I* defined depression (which it called "depressive reaction") accordingly:

> The anxiety in this reaction is allayed, and hence partially relieved, by depression and self-depreciation. The reaction is precipitated by a current situation,

frequently by some loss sustained by the patient, and is often associated with a feeling of guilt for past failures or deeds. The reaction in such cases is dependent upon the intensity of the patient's ambivalent feeling toward his loss (love, possession) as well as upon the realistic circumstances of the loss.[66]

This description relied on a few nonspecific sentences based on theoretically assumed processes, not observable symptoms. It focused solely on the psychodynamics (loss, guilt, ambivalence) that presumably led to depressive conditions but did not contain any observable criteria that would help make this particular diagnosis. This vagueness reflected the analytic and Meyerian convictions that embraced the particularity of each patient's problems.

Despite the psychoanalytic cast of depressive neurosis, the *DSM-I* maintained the time-honored distinction of what historian Edward Shorter calls the "two depressions."[67] Melancholic conditions, which were at the heart of mental illness classification systems from the Hippocratics through Kraepelin, were classified as "psychotic depressive reactions" that involve "gross misinterpretation of reality, including, at times, delusions and hallucinations."[68] Although they were called "reactions" that frequently arise in the "presence of environmental precipitating factors," their extreme presentations clearly differentiated them from psychoneurotic depressions that were defined by the "absence of malignant symptoms."[69] The two types of depression also had different treatment implications. While drugs or other somatic interventions were employed for psychotic conditions, neurotic depressions were viewed as in need of psychotherapy.[70]

Personality disorders, the fourth type of psychogenic condition, also stemmed from early modes of psychosexual development but featured minimal distress or anxiety. From the 1930s through the 1950s, many psychoanalysts placed the personality, or character, disorders at the center of their theories. They featured not so much particular symptoms as holistic character structures. Unlike psychosomatic and neurotic reactions, this class involved lifelong behavioral patterns that resisted therapeutic efforts to change them.[71]

The three personality subtypes, each of which was divided into more particular diagnoses, were basic disturbances of personality patterns, personality types that disintegrated under stressful conditions, and sociopathic personality disturbances. The latter category overtly used social values as defining elements. People who had them were "ill primarily in terms of society and of conformity with the cultural milieu, and not only in terms of personal discomfort and relations with other individuals."[72] The homosexuality diagnosis, which ultimately helped lead to the downfall of the entire *DSM-I* (and *II*) system, fell under the sexual deviation subclass of sociopathic personality disturbances. Other diagnoses in this category were "transvestism, pedophilia, fetishism and sexual sadism (including rape, sexual assault, mutilation)."[73]

The *DSM-I*'s final category, which had been at the heart of *Medical 203*, was transient situational personality disorders (TSPD). All of its four subcategories (and five additional sub-subcategories) were acute responses to overwhelming stress that did not involve a more fundamental personality disturbance. Gross stress reaction (GSR), which eventually became a model for the PTSD diagnosis that appeared in the *DSM-III*, was the most notable subcategory. Its origins clearly lay in *Medical 203*. GSR arose after either combat or civilian catastrophes such as fires, earthquakes, or explosions. In most cases, individuals who received this diagnosis were "previously more or less 'normal' persons who have experienced intense stress." The definition also noted that, with proper treatment, this reaction should "clear rapidly."[74] In addition, the class contained three adjustment reactions that arose among infants, children, and adults. By definition, all of these diagnoses were short-lived responses to stressful circumstances. If symptoms were prolonged, they indicated a different type of disorder and would not be classified as a TSPD.

In hindsight, the *DSM-I* is typically seen as reflecting a period when "there was not much interest in mental disorder classification in psychiatry, much less in psychology or anywhere else."[75] Indeed, the major history of psychoanalysis in the United States does not even mention the *DSM-I* (or *II*).[76] Yet, at the time, many viewed the manual as a sign of an *overreliance* on diagnosis. Karl Menninger used it to illustrate the "confusing complexity of American psychiatric nosology" and expressed his hope that its next revision would move it "in the direction of simplification."[77] To be sure, the *DSM-I* was far more elaborate than previous classificatory manuals were; George Raines, the chair of the committee that prepared the *DSM-I*, believed that "accurate diagnosis is the keystone of appropriate treatment and competent prognosis."[78] It provided institution-based psychiatrists and administrators the tools they required for record keeping and management while at the same time recognizing that the profession was turning away from the organic diagnoses that had dominated earlier manuals.

The etiological basis of the *DSM-I* was the major distinction between it and the *DSM* revisions from 1980 on. One of its two classes definitionally arose from damage to brain tissue, while the other, which encompassed psychogenic psychotic, psychosomatic, neurotic, personality, and stress-related disorders, was overtly psychodynamic. Another difference was that, although the *DSM-I* contained roughly 100 distinct diagnoses, few were defined explicitly. This lack of specificity in the manual's various diagnostic criteria led to a situation where "there is not even an objective method of describing or communicating clinical findings without subjective interpretation and no exact and uniform terminology which conveys precisely the same to all."[79] Nevertheless, the *DSM* remained unaltered for 16 years, when its first revision appeared in 1968.

The *DSM-II*

No group, whether inside or outside of the psychiatric profession, exerted pressure to change the *DSM-I*. Instead, the *DSM-II* emerged in 1968 from a bureaucratic effort to conform American psychiatric diagnoses to the *International Classification of Diseases*, which the World Health Organization had sponsored since 1948.[80] There is no record of any controversy, or even any debate, regarding the development of the *DSM-II*. Nevertheless, the revised manual contained some hints about the massive transformation in psychiatric diagnoses that the succeeding edition of the *DSM* was to bring about.

The *DSM-II* arose in a context where diagnosis was, at best, of marginal concern to the psychiatric profession. The dominant psychodynamically oriented group had little interest in diagnoses, although it sometimes expressed contempt for them. The small group of researchers who were intensely concerned with diagnostic issues was far outside of the field's mainstream. Robert Spitzer captured their inconsequential position best when he described APA meetings in the 1960s where "the academic psychiatrists interested in presenting their work on descriptive diagnostics would be scheduled for the final day in the late afternoon. No one would attend. Psychiatrists were not interested in diagnostic issues."[81]

Although the number of diagnoses in the *DSM-II* substantially rose, from 106 to 182, the growing number of disorders did not reflect any basic changes in philosophy.[82] It was largely the result of far greater specification in three particular classes: intellectual deficiency, alcohol and drug dependence, and behavior disorders among children and adolescents. For example, the *DSM-I* provided just two codes for mental deficiency. The revised manual offered diagnosticians six major classifications (borderline, mild, moderate, severe, profound, and unspecified), nine subcategories for each, and a plethora of further possible specifications. The first *DSM* divided addictions into alcoholism and drug addiction without additional categories. The revised manual specified four types of alcoholism and 10 types of drug dependence. It also provided a separate category for behavior disorders of childhood and adolescence that contained seven subtypes. In addition, the *DSM-II* added a new category of "special symptoms" with 10 diagnoses, including sleep and eating disorders.[83]

The second *DSM* was less etiological than the first. The committee that developed the *DSM-II* "tried to avoid terms which carry with them *implications* regarding either the nature of a disorder or its causes and has been explicit about causal assumptions when they are integral to a diagnostic concept."[84] With a few exceptions, it abandoned the Meyerian term "reaction" that implied some internal or external stress precipitated a disorder. For example, "schizophrenic reaction" became "schizophrenia." The near excision of the term "reaction" was a harbinger of the *DSM-III*'s movement from etiological to descriptive classifications.

Despite the manual's disclaimer, analytic assumptions remained, although they were less prominent in the *DSM-II* than in its predecessor. This was particularly the case for the neuroses. The introduction to this category states, "Anxiety is the chief characteristic of the neuroses. It may be felt and expressed directly, or it may be controlled unconsciously and automatically by conversion, displacement and various other psychological mechanisms." The symptoms of hysterical neurosis, for instance, were "symbolic of the underlying conflicts." Likewise, depressive neuroses were "due to an internal conflict or to an identifiable event."[85] To the extent that the *DSM-II* reflected any theoretical commitments, they were psychoanalytic in nature. Nevertheless, the new manual was quietly moving away from the pervasive psychodynamic dominance in the *DSM-I*. However, the *DSM-II* "was given hardly any attention by the analysts themselves, who did not care much about diagnosis because, as they saw it, all neurosis was caused by remarkably similar unconscious conflicts."[86]

As in the manual's first version, organic conditions were limited to those that were a direct consequence of some brain defect. The *DSM-II*, however, turned away from the strict segregation of organic from psychogenic conditions. Instead of the *DSM-I*'s "basic division" of brain-based and psychogenic conditions, the revision contained 10 general classes that did not assume a fundamental distinction between organic and other conditions.[87] This shift signaled the declining importance of clearly somatic disorders from the early 1950s to the late 1960s. Even careful observers of the changes from *DSM-I* to *DSM-II*, however, could not have foretold the revolution that the *DSM-III* would bring about.

A Nondiagnostic Culture

The *DSM-I* appeared at a time when psychiatry was entering a period of unprecedented cultural prominence. Psychoanalysis originally appealed to a relatively small group of intellectuals, bohemians, and artists who rebelled against traditional norms in the 1920s and 1930s. By the end of World War II, however, most analysts no longer challenged social norms but espoused mainstream moral, religious, and sexual practices.

The field's esteem and influence in the 16 years between 1952 and the appearance of the *DSM-II* in 1968 remains unsurpassed. Psychiatry and related professions underwent a vast expansion. When the *DSM-I* was introduced in 1952, just 1 percent of the population had seen some sort of mental health professional. By 1980, this proportion had grown tenfold.[88] Correspondingly, between 1950 and 1966 the number of psychiatrists increased from 5,500 to nearly 20,000.[89] They treated the rapidly growing number of Americans who turned to them for help solving problems in their personal relationships, raising their children, improving their sex lives, and explaining their unhappiness, anxiety, tension, fatigue, and depression.[90]

While psychiatry was an omnipresent topic in the general culture from the 1950s through the 1960s, particular diagnoses were not. Best-selling books, movies, magazines, and the mass media usually presented favorable portrayals of psychiatric theory and practice, but they didn't focus on particular *DSM* conditions.[91] The influence of psychiatry in general and psychoanalysis in particular was so great that a critic of analysis complained, "Anyone who reached adulthood prior to 1950 knows how perversely Freudian theory and practice dominated not only in the specific field of psychotherapy, but also education, jurisprudence, religion, child rearing, and art and literature, and social philosophy."[92] Accordingly, the manual's broad psychosocial character, its analytical orientation, and the central role it gave to anxiety, rather than its particular diagnoses, reflected the more general psychiatric themes that pervaded American society at the time.

Psychiatric influence peaked in the era when the *DSM-I* was the field's official classification system, but the manual itself played no role in the profession's dominance. Although developments within the psychiatric profession, particularly the movement of its practices from inpatient to outpatient settings, mandated the creation of the *DSM-I*, they did not elevate the status of diagnosis itself. Indeed, the central aspect of psychiatric diagnoses at the time was their very *unimportance*. The manual went through 12 printings by 1958, but there is no evidence that psychiatrists actually made much use of it.[93] Likewise, although the *DSM-II* sold well, like the first edition, it was culturally invisible.

Outside of mental institutions, which themselves began a precipitous decline in the mid-1950s, it is hard to see what the first two *DSMs could* have been used for. Clinicians did not require specific diagnoses to be reimbursed for their services from third-party payers, which barely existed in the 1950s. The discoverer of the tranquilizer Miltown, Frank Berger, assessed the role of diagnosis at the time:

> My feeling was that most people we saw had really no psychiatric disorders. They were people, in my opinion, with problems of living, people who did not get on with their spouses, did not get on with their children, did not get on with their boss, and had not been taught, had not been educated, had not been prepared to handle all these crises of life. So they got stressed, broke down, and had to see a doctor, and the doctor did not know what to do. So he put one of the psychiatric names on them.[94]

Patients, too, had no need for a particular diagnosis to obtain prescriptions or to enter psychotherapy. Psychiatric researchers, of whom there were very few, studied broad categories such as "stress"; specific diagnoses had not yet become central to the research community. The National Institute of Mental Health, the major funder of psychiatric research and education, was more interested in addressing broad social problems than in facilitating better classification systems. Psychoanalysts, who emphasized broad unconscious conflicts unconnected to any specific effects that underlay a huge array of symptoms, dominated psychiatry departments in medical schools. There was, in short, no need for anyone to use any but the most general classifications.

Little information exists about how psychiatrists and other physicians actually employed the *DSM-I* and *II*. The National Disease and Therapeutic Index, a commercial database initiated in 1958 by IMS Health, provides one of the few sources about the application of psychiatric diagnoses at the time. Its data stem from a sample of more than 3,000 physicians, including psychiatrists, that was meant to represent all medical practices in the United States.[95] The results indicated that nearly 44 million visits were made to physicians for all sorts of mental disorders in 1962. When psychiatric outpatients did receive labels, they were usually of some neurosis or personality disorder, which accounted for more than 36 million visits. The 11.6 million anxiety reactions were the most common single diagnosis. In contrast, all the psychoses accounted for a little over 7 million visits. Depression comprised just 4 million.[96]

Americans turned to psychiatrists to explore their inner selves, explain their unhappiness, fix their intimate relationships, and improve their lives. They did not, however, expect to receive particular diagnoses of their problems, and they typically did not get them. Sociologist Charles Kadushin's study of 1,500 outpatients in psychiatric clinics in New York City in 1959–60 indicated that the most common complaints involved general anxiety and dissatisfaction, sexual problems, and interpersonal difficulties: "The empirically derived clustering here presented covers a wide range of problems; indeed it is one way of grouping all of life," Kadushin summarized.[97]

Google nGram analysis also reveals the inconspicuousness of the *DSM-I* and *II*: the term "DSM" was all but nonexistent in books published before the late 1970s. "The DSM was rarely mentioned from 1960 to the late 1970s, befitting its neglect among professionals, scientists, and certainly the general public," sociologist Jason Schnittker concludes.[98] In contrast, nGram analysis indicates a sharp rise in the terms "neurosis" and "psychosis" from the late 1930s through the late 1950s, when mentions of these general labels started to decline. "Anxiety" was by far the most commonly employed psychiatric term from the 1950s through 1980, when its use leveled. Neither specific diagnoses nor the *DSM* itself were aspects of the pervasive cultural concern with mental illness during the postwar period.

Psychotropic Drugs

The *DSM-I* and *DSM-II* were, at best, minor conduits for conceptions of mental illness at the time. Their broad scope mirrored but did not shape popular views of psychiatric disorders. Instead, promotions for the psychotropic drugs that became enormously popular in the mid-1950s provide the best window onto common notions of patient conditions. Meprobamate, usually marketed under the trade names Miltown and Equinal, entered the pharmaceutical arena in 1955. It was a muscle relaxant with mildly sedative properties that was called a "minor tranquilizer." The minor tranquilizers eased tension and anxiety while allowing people to maintain their everyday activities. Drugs such as Miltown were prescribed for "perfectly normal people who need temporary help." They were called "happy pills," "calming pills," or "emotional aspirin," names that carried no diagnostic connotations.[99]

Miltown quickly became the most widely prescribed drug of any sort in history: within a year of its introduction, 5 percent of all Americans had used it. By 1957 the "minor tranquilizers accounted for a staggering one-third of all prescriptions." These drugs were especially suitable for patients of general physicians and internists, who wrote about four out of five prescriptions for the minor tranquilizers. By 1960 about three-quarters of physicians prescribed meprobamate to some of their patients. "Meprobamate opened up the question of the mass treatment of nervous problems found in the community," psychopharmacologist David Healy concludes.[100]

Drug companies advertised their products widely in psychiatric and medical journals. Miltown's purported uses did not fall into even the broadest *DSM-I* or *II* diagnostic categories but spanned a huge array of illnesses, "nerves," and problems of living. Ads accordingly emphasized their applicability to capacious states of "mental stress" and "tension" in "the average patient in everyday practice."[101] One that appeared in the July 1964 issue of the *American Journal of Psychiatry* touted Miltown's use for a dizzying array of issues including problem children, alcoholics, insomniacs, senility, tenseness and nervousness, and anxious depression. "Virtually any of your patients, regardless of age or disorder can be given the drug with confidence as an adjunct to psychotherapy or any other therapy you may employ," the ad announced (figure 2.1). In short, psychoses aside, general physicians and psychiatrists could use Miltown for the whole range of disturbances they saw.

Other ads declared, "For your patients, Miltown promptly checks emotional and muscular tension. Thus, you will make it easier for them to lead a normal family life and to carry on their usual work." An ad indicated the tranquilizer Ultram for "the patient with vague symptoms, nervous and distressed under the burden of unsolved problems, finds release from anxiety and restoration of emotional composure."[102] Historian David Herzberg observes, "The ads translated psychodynamic psychiatry's obsession with anxiety into simple life problems that physicians confronted every day in their offices."[103]

Although drug companies were allowed to advertise their products only to physicians at the time, the minor tranquillizers were widely hailed in mass circulation magazines and newspapers, which encouraged their use for dealing with work, family, marriage, and sexual problems.[104] The popular media was filled with stories about them. An article in *Cosmopolitan* touted Miltown for such varied symptoms as stomach distress, the "blues," fatigue, lack of "social ease," insomnia, and skin problems.[105] Other pieces portrayed the minor tranquillizers as remedies for the stresses of daily life rather than for any kind of disease. One newspaper reported, "A Louisville truck driver was the excitable type and would get 'mad' when another driver cut in front of him. He would cuss out the other driver and perhaps run into his car. But since he has been taking the new drug, he remains calm and hasn't had an accident." Another article included a wife's description of the influence the drug had on her spouse: "My husband is on it now. He used to

BELONGS IN EVERY PRACTICE OF PSYCHIATRY

nervous patient

The agitated senile patient

Anxious depression

It's versatile: The years have proved that 'Miltown' (meprobamate) is the one tranquilizer that is helpful in almost every aspect of psychiatric practice. Virtually any of your patients, regardless of age or disorder, can be given the drug with confidence as an adjunct to psychotherapy or any other therapy you may employ.

Outstanding record of safety: Over eight years of clinical use among millions of patients throughout the world – plus more than 1500 published reports covering the use of the drug in almost every field of medicine – support your prescriptions for 'Miltown' (meprobamate). This is why it "belongs in every practice."

Dependable: 'Miltown' (meprobamate) is an established drug. There are no surprises in store for you or your patient. You can depend on it to help your patients through periods of emotional distress – and to help maintain their emotional stability.

Easy to use: Because 'Miltown' (meprobamate) is compatible with every other psychotherapeutic measure, you'll find it fits in easily with any treatment you are now using. It will not, therefore, complicate treatment of patients seen in psychiatric practice.

the original brand of meprobamate
Miltown

The alcoholic

Figure 2.1. Tranquilizer advertisements presented them as treatments for an extremely wide range of problems and types of people. *Psychosomatic Medicine,* 26(2), 1964.

be very nervous, really just miserable. Now he doesn't get mad as quick or stay mad as long. He has no energy, of course."[106]

Feminist Betty Friedan's turn of phrase, "the problem that has no name," perhaps best captured the nondiagnostic nature of these drugs.[107] Friedan associated this nameless problem, which featured "mild, undiagnosable symptoms . . .

malaise, nervousness, and fatigue" with the social role of housewives, who faced unfulfilling lives centered on housework, cooking, and taking care of children. As a consultant to the Food and Drug Administration claimed, "We are faced with the possible utilization of psychoactive drugs by a major portion of the population in what might be considered problems of daily living."[108]

Although Friedan and many others criticized the use of tranquilizers for controlling women's distress, others were more favorable. Anthropologist Margaret Mead surprised senators at a congressional hearing inquiring into the misuse of tranquilizers when she proclaimed:

> I do believe it is worthwhile to avoid the stress that comes when the plumbing breaks down and both cars are broken and you can't find your husband to telephone him, and the child in nursery school, three children in nursery school, you were going to pick up 15 miles somewhere else, if a pill will permit you not to burst into tears under these circumstances but go next door and borrow another car, I think it is a good idea.[109]

In contrast to the critics, Mead saw little downside to using a pill to ease the psychic distress that arose from women's problems of living.

In the 1960s the benzodiazepines Librium and Valium, which were also marketed for the relief of "anxiety and tension," displaced Miltown as the next drug sensations. Like Miltown, these drugs were hugely popular. In 1971 an article in *Fortune* magazine called Valium and Librium "the greatest commercial successes in the history of prescription drugs." At the time, about 15 percent of all Americans, most of whom were women, reported using Valium or a related pill in the past year. By the late 1960s and early 1970s, Valium and Librium were the first- and third-most-prescribed drugs of any type in the United States. Perhaps most astonishing, half of all Americans said they had used a tranquilizer at least once.[110]

The development of Librium and Valium intensified drug promotion. "The April 1978 issue of the *American Journal of Psychiatry*," historian Andrea Tone reports, "contained an astounding sixty-four pages of advertisements."[111] The benzodiazepines were heralded for their role in dealing with stressors both large and small. Accordingly, ads for Librium touted its ability to assist people in adjusting to the turmoil of the 1960s and its "growing role in helping mankind to meet the challenge of a changing world." At the other end of the life event spectrum, Valium was promoted as "the 'daytime sedative' for everyday situational stress. When stress is situational—environmental pressure, worry over illness—the treatment often calls for an anxiety-allaying agent." The widespread acclaim the mass media accorded these drugs expanded the already vague notions of "anxiety" and "neurosis" to incorporate a vast array of problems of living (figure 2.2).[112]

General physicians prescribed the great majority of tranquilizing drugs. "What illnesses are being treated?" prominent pharmacologist Barry Blackwell asked. "Most of what primary care physicians see, they label 'anxiety.'" Finer diagnostic distinctions were unnecessary. In the 1950s and 1960s a diagnosis of some psychiatric disorder was present in a minority of prescriptions: "Only one-third of the prescriptions for the minor tranquilizers were written for diagnoses of mental, psychoneurotic, and personality disorders," pharmacologist Mickey Smith indicates. During the *DSM-I and II* era, particular diagnoses had little role to play in either the promotion of psychotropic drugs, the psychiatric profession, or the broader society. "Like their predecessors, Librium and Valium quickly gained a reputation for combating a breathtakingly wide range of illnesses," Herzberg notes.[113]

The phenothiazine Thorazine entered a very different drug market around the same time as Miltown. Thorazine, which was originally used to sedate patients

35, single and psychoneurotic

The purser on her cruise ship took the last snapshot of Jan. You probably see many such Jans in your practice. The unmarried with low self-esteem. Jan never found a man to measure up to her father. Now she realizes she's in a losing pattern—and that she may *never* marry.

Valium (diazepam) can be a useful adjunct in the therapy of the tense, over anxious patient who has a neurotic sense of failure, guilt or loss. Over the years, Valium has proven its value in the relief of psychoneurotic states—anxiety, apprehension, agitation, alone or with depressive symptoms.

Valium 10-mg tablets help relieve the emotional "storms" of psychoneurotic tension and the depressive symptoms that can go hand-in-hand with it. Valium 2-mg or 5-mg tablets, *t.i.d.* or *q.i.d.*, are usually sufficient for milder tension and anxiety states. An *h.s.* dose added to the *t.i.d.* dosage often facilitates a good night's rest.

Valium® (diazepam)
for psychoneurotic states manifested by psychic tension and depressive symptoms

 Roche

Before prescribing, please consult complete product information, a summary of which follows:
Indications: Tension and anxiety states; somatic complaints which are concomitants of emotional factors; psychoneurotic states manifested by tension, anxiety, apprehension, fatigue, depressive symptoms or agitation; acute agitation, tremor, delirium tremens and hallucinosis due to acute alcohol withdrawal; adjunctively in skeletal muscle spasm due to reflex spasm to local pathology, spasticity caused by upper motor neuron disorders, athetosis, stiff-man syndrome, convulsive disorders (not for sole therapy).
Contraindicated: Known hypersensitivity to the drug. Children under 6 months of age. Acute narrow angle glaucoma.
Warnings: Not of value in psychotic patients. Caution against hazardous occupations requiring complete mental alertness. When used adjunctively in convulsive disorders, possibility of increase in frequency and/or severity of grand mal seizures may require increased dosage of standard anticonvulsant medication; abrupt withdrawal may be associated with temporary increase in frequency and/or severity of seizures. Advise against simultaneous ingestion of alcohol and other CNS depressants. Withdrawal symptoms have occurred following abrupt discontinuance. Keep addiction-prone individuals under careful surveillance because of their predisposition to habituation and dependence. In pregnancy, lactation or women of childbearing age, weigh potential benefit against possible hazard.
Precautions: If combined with other psychotropics or anticonvulsants, consider carefully pharmacology of agents employed. Usual dosage indicated in patients severely depressed, or with latent depression, or with suicidal tendencies. Observe usual precautions in impaired renal or hepatic function. Limit dosage to smallest effective amount in elderly and debilitated to preclude ataxia or oversedation.
Side Effects: Drowsiness, confusion, diplopia, hypotension, changes in libido, nausea, fatigue, depression, dysarthria, jaundice, skin rash, ataxia, constipation, headache, incontinence, changes in salivation, slurred speech, tremor, vertigo, urinary retention, blurred vision. Paradoxical reactions such as acute hyperexcited states, anxiety, hallucinations, increased muscle spasticity, insomnia, rage, sleep disturbances, stimulation, have been reported; should these occur, discontinue drug. Isolated reports of neutropenia, jaundice; periodic blood counts and liver function tests advisable during long-term therapy.

Figure 2.2. Advertisements commonly employed undesirable social situations such as being a single woman as reasons for taking tranquilizers. *Archives of General Psychiatry*, 22, 1970.

undergoing surgery, was found to calm institutionalized patients, restore their sense of reality, make them accessible to treatment, and, often, allow them to be released into the community. Within a year of its introduction, psychiatrists had written more than 2 million prescriptions for the drug. Because of the distinct psychotic and neurotic populations they were used with, Thorazine was called a "major tranquilizer" in contrast to the "minor tranquilizer" Miltown. Hospital-based psychiatrists prescribed the former while general practitioners or, less often, psychiatrists prescribed the latter as adjuncts to verbal explorations rather than as replacements for them.[114] The basic psychotic/neurotic distinction sufficed to divide drugs into the major and minor tranquilizers. The mania for the tranquilizers that marked American society during the 1950s and 1960s did not require sharp diagnostic distinctions, or even any diagnoses at all.

The lack of attention to the *DSM-I* and *II* classifications was unsurprising. These manuals generated no interest in the broader culture and not much more in the psychiatric profession itself. "As a minor desk reference, DSM-II was neither prominent nor controversial among practicing clinicians. It did not impose on their treatment plans or restrict their professional judgment. DSM-II had few defenders or spokespersons because no one sensed a need to protect it," according to Stuart Kirk and Herb Kutchins.[115] Although psychiatry, psychoanalysis, and the tranquilizing drugs suffused American culture in the 1950s and 1960s, the *DSM* did not. In a short time, this situation would dramatically change.

Troubles Begin

The *DSM* finally entered the public spotlight in the early 1970s when a sharp conflict arose over the manual's definition of homosexuality.[116] For millennia, religious and cultural attitudes had considered this behavior to be a moral abomination. Although Freud himself had a benevolent view of homosexuality and did not consider it to be a mental disorder, after his death, many analysts came to believe that it was a pathology arising from unresolved unconscious infantile conflicts with rejecting fathers and overprotective mothers.[117] They considered viewing homosexuality as a neurosis or personality disorder to be far more enlightened than seeing it as a sin or a crime. It was the first condition the *DSM-II* listed under the category of sexual deviations intended for use with "individuals whose sexual interests are directed primarily toward objects other than people of the opposite sex, toward sexual acts not usually associated with coitus, or toward coitus performed under bizarre circumstances as in necrophilia, pedophilia, sexual sadism, and fetishism. Even though many find their practices distasteful, they remain unable to substitute normal sexual behavior for them."[118]

Forthright advocacy of considering homosexuality to be an alternative sexual orientation rather than a mental pathology did not coalesce until the late 1960s. In a social climate where the counterculture and feminists challenged traditional definitions of "normal" and "abnormal," gay activists began to confront psychiatry's classification of homosexuality as a mental disorder. They argued that this behavior was no more pathological than heterosexuality. Militants challenged the psychiatric definition of homosexuality at a 1972 meeting of behavior therapists that Spitzer was attending. He invited them to participate in a panel that he would arrange about this issue at the APA's annual meeting the following year. On one side would be psychoanalysts who were not just committed to the position that homosexuality was a mental disorder but also appalled that any outside group could dictate what conditions psychiatry should or should not label mental disorders. On the other side, psychiatrists who were dedicated to eliminating homosexuality from the diagnostic manual would speak. The debate was highly publicized in advance and attracted an audience of more than 1,000 psychiatrists as well as vast attention in the national media.

As an outgrowth of this session, Spitzer developed and circulated a definition of homosexuality that stated it should not be considered a mental disorder unless it was also accompanied by intense distress. That is, homosexuality in itself was not a psychiatric disorder but would remain in the *DSM-II* as a "sexual orientation disturbance" that applied only to those who wanted to change their sexual orientation. Neither the activists nor the analysts were entirely satisfied with this compromise, but both sides grudgingly accepted it. The new diagnosis was put to a vote of the relevant APA council, which unanimously approved it. The APA Assembly and Trustees then confirmed the council's decision.[119]

The notion that a vote could decide whether some condition was a disease was a severe blow to psychiatry's prestige. Front-page headlines in major media outlets trumpeted, "Doctors Rule Homosexuals Not Abnormal," "Victory for Homosexuals," and "Psychiatrists in a Shift, Declare Homosexuality No Mental Illness."[120] The seventh printing of the *DSM-II* (1974) incorporated a black-boxed "special note" indicating that the APA had substituted a new category of sexual orientation disturbance for the now-excised homosexuality diagnosis.[121] The intense controversy over homosexuality was an augur of conflicts that would mark the development of the *DSM-III*. In particular, the fracas proved to be the first step in the wholesale slaughter of psychoanalytic perspectives that the next version of the *DSM* would bring about.

Toward *DSM-III*

It could not have been known at the time, but the major link between the *DSM-II* and the subsequent *DSM-III* diagnostic revolution was the participation of one individual—Robert Spitzer—in the revision process.[122] As a precocious teenager, Spitzer had entered Reichian psychoanalysis, which promoted the enhancement of sexual powers. He went on to become a psychoanalyst himself, although he quickly became disillusioned with this method. Instead, Spitzer allied with the small number of research-oriented psychiatrists who were particularly interested in designing standardized diagnostic classifications. He was perhaps the only member of this group who was also heavily involved with the APA's revision of *DSM-I*. The organization had invited a Trojan horse into the diagnostic process. Spitzer was a consultant to the committee that developed the *DSM-II* and wrote the only official APA publication about the new manual. This article, which also appeared as the final section of the manual itself, discussed the differences between the first two versions of the *DSM*. In retrospect, it pointed to the new directions that the *DSM-III* would take in the following decade.[123]

Spitzer emphasized the "striking difference" that the elimination of the term "reaction" from most of the *DSM-II* diagnoses created between the first two manuals. This move eliminated a label that had strong Meyerian and analytic causal implications and provided the first indication that the *DSM* diagnostic system was moving in a more purely descriptive direction. Another change that anticipated future *DSM*s was the elimination of the sharp distinction between organic and all other mental disorders. Instead, organic brain syndromes were just 1 of 10 major classes in the *DSM-II*. A third change presaging the next manual was the loosening of the *DSM-I* hierarchical model in favor of encouraging multiple diagnoses in the same patient. This move would eventually lead to the focus on comorbidity when the *DSM-III* greatly amplified the emphasis on splitting rather than lumping diagnoses. Spitzer also pointed attention to the greater specificity in several of the new manual's major diagnostic categories, including schizophrenia, sexual deviation, and alcohol and drug dependence. Finally, the insertion of a new class of behavior disorders of children and adolescence foreshadowed the far greater attention future *DSM*s would pay to mental disorders among young people.

These changes, at least partly the result of Spitzer's instigation, were virtually unnoticed at the time. They were, however, important portents of the empirically based, theoretically neutral diagnostic system that would emerge once Spitzer took control of the revolution in psychiatric diagnosis that began in the 1970s. He uniquely understood that leadership of the *DSM* revision potentially had the power to radically change the nature of the psychiatric profession itself. Gaining control over the *DSM-III* process would make Spitzer the most influential American psychiatrist of the latter half of the twentieth century.

| The Path to a Diagnostic
Revolution

The *DSM-I* was well suited for the needs of psychiatrists and other clinicians in
the 1950s and much of the 1960s. In that period specific diagnoses served
no professional, economic, or cultural purposes. Psychiatric researchers, who
did require well-defined diagnostic criteria, were few in number and not highly
valued in the profession. Third parties who might have demanded diagnoses did
not have major roles in the economic aspects of therapeutic encounters: hospital-
based psychiatrists were salaried while patients themselves reimbursed outpatient therapists. By the time the *DSM-II* became available in 1968, all these
factors were in flux.

Pressures for Specific Diagnoses Emerge

Although psychiatry was highly visible in the general culture, the *DSM* was not.
Particular diagnoses had few uses, rarely influenced treatment, and were not
highly valued. "Interest in psychiatric diagnosis in the 1960s was probably at an
all-time low in the United States," psychiatrist Andrew Skodol observed.[1] However, the rise of a biomedically oriented group of psychiatrists and demands from
government agencies, third party insurers, and the drug industry eventually produced an entirely new diagnostic system.

BIOLOGICAL PSYCHIATRY DEVELOPS

A number of factors, both within and outside of psychiatry, created intense demands on the field to discard the *DSM-II* and replace it with a completely different type of classification. Inside the profession, a subspecialty was arising that was far more focused on diagnosis than its psychodynamic counterpart was. Before the 1950s virtually nothing was known about brain chemistry. The decoding of DNA in 1953 generated intense new interest in this organ. From that time onward researchers strove to identify specific neural mechanisms associated with various mental disorders. Although the idea that particular brain malfunctions were responsible for mental pathology revived a dominant strain of thinking in nineteenth-century psychiatry and neurology, the biological view seemed genuinely new at the time.

As psychopharmacologists began to serendipitously discover a variety of psychotropic drugs, including phenothiazine, imipramine, and the monoamine oxidase inhibitors (MAOIs) in the 1950s, researchers became attracted to the notion that drugs targeted at specific brain regions could aid in the diagnostic process itself. Psychiatrist Donald Klein proposed the idea of "psychopharmacological dissection" to show how the close observation of the varying effects drugs produced among different patients could help diagnose what specific conditions they had in the first place.[2] The antidepressant imipramine provided an example: "All research workers who used this drug initially were unanimous in the conclusion that, unlike many other psychopharmacological agents, it specifically affects depressive conditions and has very little effect on paranoid states or disturbed behavior, particularly in schizophrenics."[3] The conviction developed that imipramine or an MAOI should be directed at depression, phenothiazine at schizophrenia, and lithium at manic depression.

The narrative emerging from psychopharmacological research thoroughly contrasted with the analytic focus on uncovering unconscious conflicts from the distant past and using the therapeutic relationship to resolve them. A distinctly nonanalytic conception of mental illness also started to penetrate US culture in the mid-1950s when popular media began to pay attention to biologically based research and treatments. In 1955, Congress passed the Mental Health Study Act, which allocated $2 million annually to the National Institute of Mental Health for psychopharmacological research.[4]

By the mid-1960s, biological psychiatry had developed to the point that it presented a genuine challenge to the psychoanalysts who dominated the profession.

Adherents of the new perspective spoke an entirely different language than that of the analysts, one grounded in neurochemistry, double-blind research trials, and the primacy of drug therapies. They openly challenged the analysts' casual approach to research. One of its leaders, Gerald Klerman, called for far more rigorous research standards, stating that psychotherapy and psychoanalysis were consuming ever-larger amounts of the health budget, with little evidence that they worked for anything in particular.[5]

Most biologically oriented researchers worked in academic or laboratory, not clinical, settings. They were usually tied to medical schools, which took for granted that any legitimate specialty, including psychiatry, would use traditional scientific methods, study specific diseases, and employ testable therapies. Accordingly, throughout the 1960s and 1970s, the number of psychiatry departments with analytically oriented chairs declined while those with empirically minded ones grew. Nevertheless, analysts remained powerful within the psychiatric profession: as late as 1973 about half of psychiatrists specialized in psychoanalysis.[6]

FEDERAL AGENCIES DEMAND DIAGNOSES

Changing approaches to mental illness within the federal government were also central in elevating the need for a new diagnostic system that was grounded in specificity. Two agencies, the National Institute of Mental Health and the Food and Drug Administration pushed the field to place greater emphasis on diagnoses. The NIMH was the major source of funding for psychiatric research and education. At the same time research-oriented psychiatrists were turning away from the psychosocial model, the NIMH confronted an intense political backlash over its initial focus on social issues.[7] Since the mid-1960s, this agency's activist policies and resulting perception that it did not deal with true diseases led Congress to consistently cut the NIMH budget and, consequently, funding for psychiatric researchers. A new diagnostic system grounded in disease-like entities was hoped to be the perfect antidote for the now-toxic focus on social problems. The NIMH became an important partner of the American Psychiatric Association in sponsoring a new biomedically oriented DSM that would serve their mutual interests.

The FDA was a second federal agency that turned psychiatry toward the study of specific diseases. In the early 1960s, a public uproar resulted when the sedative drug thalidomide was found to produce severe birth defects in the children of mothers who had taken it for relief of morning sickness. In response to this crisis, in 1962 Congress passed the Kefauver-Harris drug amendments that directed the FDA to scrutinize more tightly which drugs could come to market and how they should be used. The agency began to mandate randomized, placebo-controlled, and double-blind trials as necessary demonstrations of drug efficacy. This directive stimulated the need for a new diagnostic system because such procedures required members of both the treated and the control groups to have the same disease condition. Congress and the FDA made no distinction between psychiatric and medical conditions, assuming that both could be defined in highly specific ways. Yet, the extant *DSM* criteria were too vague and in many cases too infused with psychodynamic assumptions to serve as the basis for defining the targets of particular drugs. "After 1962," psychopharmacologist David Healy notes, "a standardization of diagnostic practice was all but inevitable."[8]

Paradoxically, the highly publicized anti-tranquilizer fervor in the early 1970s also gave momentum to the drive toward diagnostic specificity. This feminist-led movement claimed that these drugs did not treat genuine diseases but were promoted and used to keep women in subordinate social roles. Media portrayals of the tranquilizers, which were highly favorable before this time, turned sharply against them. The increasingly negative publicity surrounding these extraordinarily widely used drugs portrayed them as potentially addictive, overprescribed, and aimed at problems of living rather than genuine diseases.[9]

The FDA began to insist that legitimate use of psychotropic drugs must be limited to the treatment of particular disease states. In 1971 its commissioner, Charles Edwards, told a congressional subcommittee investigating the drug industry that Librium was "not approved for treating an everyday situation or ordinary life problem. What has happened is the advertisers, the promoters of these drugs, have taken these symptoms out of context and applied them to the day-to-day situations that we are all confronted with. As a result they have developed this totally misleading advertising" (figure 3.1). In that year, the FDA sent a letter to pharmaceutical companies ordering them to stop promoting their products as ways to ease everyday life stressors or to treat such nonspecific conditions as "tension," "nervousness," or "anxiety." At the same time it notified all sponsors of new psychotropic drugs that they could gain approval only "for the treatment of clinically significant anxiety, depression, and/or other mental conditions."[10]

In 1975 the Drug Enforcement Administration classified Librium and Valium as Schedule IV controlled substances, which required tighter reporting standards, imposed limits on prescription refills, and mandated labeling that warned of the drugs' potential for abuse. Four years later, Valium packaging was changed to read, "Valium is indicated for the symptomatic relief of anxiety and tension associated with anxiety disorders, transient situational disturbances and functional or

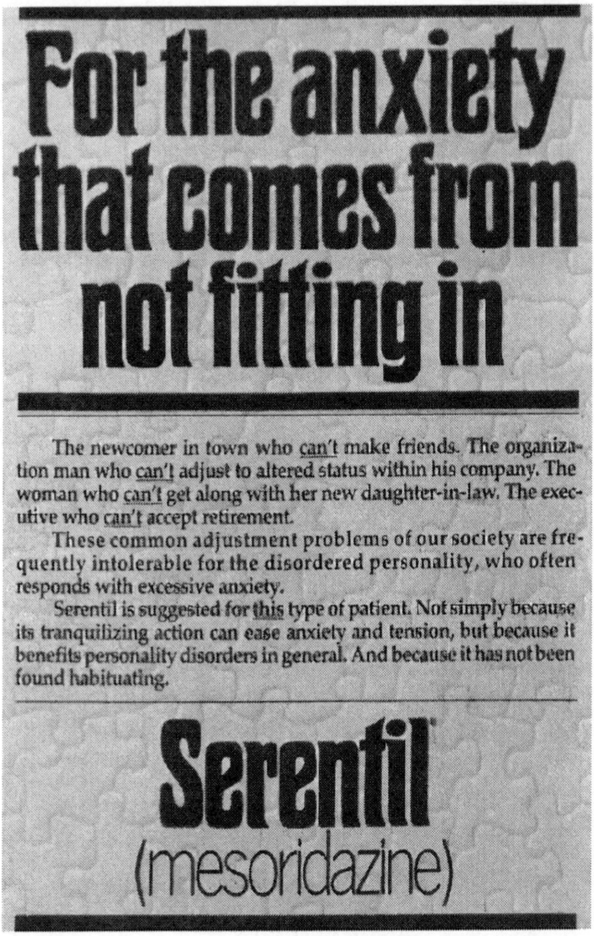

Figure 3.1. This advertisement for Serentil led the Food and Drug Administration to ban the promotion of drugs solely for treating problems of living. *Medical World News,* December 11, 1970.

organic disorders." The following year the FDA mandated that manufacturers state that "anxiety or tension associated with the stress of everyday life usually does not require treatment with an anxiolytic." Sales of the minor tranquilizers began to decline: between 1973 and 1981, benzodiazepine use dropped by 33 percent. By 1983 the best-selling tranquilizer was only the 32nd-most-prescribed drug, a precipitous fall from just a decade before.[11] "Once greeted with enthusiasm in the patients' world as an alternative to the barbiturates, Librium and Valium now seemed like a curse devised by the wicked drug companies to plague the walking wounded," historian Edward Shorter observes.[12]

As later developments would show, almost all psychotropic drugs do not work specifically for particular *DSM* mental disorders. Nevertheless, declining tranquilizer sales, coupled with congressional pressure and the resulting FDA mandate, provided powerful reasons for the pharmaceutical industry to market its products as remedies for specific diseases. Biological psychiatrists, pharmaceutical companies, the FDA, and the NIMH all began to push a narrative centered on the specificity of diagnostic categories. "During the 1970s the major psychiatric disorders became defined as disorders of single neurotransmitter systems and their receptors, with depression being a catecholamine disorder, anxiety a 5HT disorder, dementia a cholinergic disorder, and schizophrenia a dopamine disorder," Healy explains.[13]

A CHANGING ENVIRONMENT

Other powerful sources of dissatisfaction were also forcing psychiatry to create a more medical-seeming diagnostic classification system. The field was losing its allure to the general culture as the favorable postwar portrayals of psychiatrists and psychiatry turned sharply critical. For example, the heroic depictions of the profession in movies during the 1950s and early 1960s became highly negative.[14] Well-known attacks from figures such as psychiatrist Thomas Szasz and, later, psychologist David Rosenhan centered on the profession's inability to define even its most basic concepts, such as "mental illness" or "schizophrenia."[15] Rosenhan, for example, asserted that "the view has grown that psychological categorization of mental illness is useless at best and downright harmful, misleading, and pejorative at worst."[16] Widespread anti-psychiatric cultural influences pushed the discipline to develop a more medical-like diagnostic system to counter these critiques.

Within psychiatry, the relatively small group of researchers was not the only part of the field struggling with the lack of well-defined diagnoses. The far larger numbers of clinicians were facing their own diagnostic crisis. Unlike researchers, who were funded through their academic positions and research grants, practitioners earned their livings from the therapy they provided. Before the 1960s, most patients paid for their own treatments. In that decade, private and public third-party reimbursement began to expand rapidly. Between the mid-1960s and 1980, the percentage of patients using insurance to pay for outpatient psychotherapy rose from 38 to 68 percent.[17] While patients had little concern with particular diagnostic categories, the third parties that increasingly paid for their treatment were coming to demand that it involve genuine diseases and not problems in living. Insurers were willing to fund only a specific number of visits for well-defined problems and cast an especially sharp eye on what could often be interminable psychoanalytic sessions. The nebulous *DSM-II* conditions poorly fit an insurance logic that would reimburse the treatment of only discrete diseases.

The skyrocketing growth of the mental health professions during the *DSM-I* and *II* era was another factor pushing psychiatry to develop a more medical-seeming diagnostic system. The success of analytic culture had spawned a demand for treatment in the postwar period that exceeded supply. Between the end of World War II and the mid-1970s, the proportion of Americans who used some kind of mental health professional nearly doubled, from 14 to 26 percent.[18] In 1947 there were only 4,700 psychiatrists; by 1976 this number had swelled to more than 28,000. From 1964 to 1970 alone, the number of psychiatrists increased by 36 percent, compared to 9 percent among all physicians. Outpatient episodes exploded from just 379,000 in 1955 to 4.6 million in 1975. Yet while the number of psychiatrists grew rapidly in this period, the number of clinical psychologists, social workers, and marriage and family counselors, among others, expanded at even higher rates. The overall number of mental health professionals rose from 23,000 in 1947 to more than 104,000 in 1976.[19]

Psychiatrists, who charged considerably higher fees than their competitors did, needed some way to differentiate themselves from nonmedical clinicians. "Apart from their training in medicine," one leading psychiatrist asserted, "psychiatrists have nothing unique to offer that cannot be provided by psychologists, the clergy, or lay psychotherapists."[20] A clearly medical diagnostic system would abet this important goal.

Pressures from both within and outside of the psychiatric profession, therefore, led to conditions ripe for a revolution in the *DSM*. The growing chorus that promoted specific diagnoses found its perfect conductor: Robert Spitzer. As the previous chapter noted, Spitzer had assumed an important, if inconspicuous, role in the *DSM-II* revision process. In 1973 he had also represented the APA during the crisis over the manual's diagnosis of homosexuality. Because no one except for Spitzer and a few like-minded research psychiatrists paid much attention to the *DSM*, there was no controversy when the APA appointed him to lead the process to revise the *DSM-II*. "When Bob was appointed to the *DSM-III*," his colleague Donald Klein observed, "the job was of no consequence. In fact, one of the reasons Bob got the job was that it wasn't considered that important. The vast majority of psychiatrists, or for that matter the A.P.A., didn't expect anything to come from it."[21] This decision, which was virtually unnoticed at the time, created a perfect match between a fervent believer in the importance of specific diagnoses and a profession that desperately needed them. As Spitzer wrote in 1976, "DSM-III will not only be a radical departure from DSM-II—it will be revolutionary!"[22] His comment was prescient: the *DSM-III* thoroughly transformed the theory and practice of psychiatry and other mental health professions.

Developing the *DSM-III*

As the official diagnostic manual of the APA, the *DSM* represents the profession's view of what counts as a mental disorder. Between 1952, when the organization published the first version, and the early 1970s, psychiatry underwent massive changes. When the first manual arose, analytic psychiatrists, who took a casual view of measurement, were in the process of taking power from hospital-based practitioners. Neither group was vested in diagnostic issues. By the 1970s, the APA consisted of constituencies with highly divergent assumptions and interests, including the dwindling group of institutional psychiatrists, clinicians holding a variety of theoretical perspectives, and researchers who were highly attuned to diagnostic issues. The process of creating the *DSM-III* featured the striking disparities in outlook that had developed between researchers and clinicians.

Spitzer's mandate was to develop a new diagnostic system that clinicians and researchers alike could use. This posed a formidable challenge. The first two *DSM*s, to the extent that anyone used them, were designed for practitioners; they were essentially useless for research purposes. Most of their global descriptions lacked specific criteria, so psychiatrists could interpret them in many ways. The result was widespread disagreement on how to apply diagnostic labels to individual cases. This was not a problem when the APA designed the *DSM-I* and *II* because researchers were not a significant constituency in the 1950s and were just starting to gain professional influence in the 1960s.

By the 1970s, however, researchers had gained powerful positions in the APA, and the organization was becoming far more concerned with issues surrounding evidence-based medicine. Psychiatry departments in medical schools, the NIMH, and the broader culture were also shedding their infatuation with analysis and activist orientations in favor of more traditional scientific approaches to mental disorder. "The history and development of the DSM-III is a story about the changing power base, as well as the changing knowledge base, within American psychiatry. Clinicians were replaced by biomedical investigators as the most influential voices in the field," psychiatrist Mitchell Wilson observes.[23]

CLINICIANS AND RESEARCHERS

The conflicting perspectives and interests of clinicians and researchers became the central dynamic in the development of the *DSM-III*. Clinical work involved the particular qualities of each client. As one New York analyst put it, clinicians dealt with "interweaving disturbed functions embedded in . . . the very fabric of the patient's existence."[24] For analytically oriented clinicians, observable symptoms actually posed barriers to obtaining deeper understandings of the meaning of a patient's problems. They were far more concerned with answering Karl Menninger's question "What is behind the symptom?" than with determining what specific disorder their clients might have.[25]

In sharp contrast to clinicians, researchers and their allies in federal agencies and drug companies require standardized categories that are measured in the same ways across different individuals and different research sites. The particularities of individuals that are of primary interest to clinicians present obstacles to these aims. Researchers study mental disorders, not the people who have them. "A common misconception," the *DSM-III*'s introduction stated, "is that a classification of mental disorders classifies individuals, when actually what are being classified are disorders that individuals have."[26]

Another central distinction between practitioners and investigators is the role of clinical intuition. The particular insights and relationships therapists have with their patients are essential aspects of clinical practice. Indeed, the process of "transference," which involves the specific psychological dynamics in therapeutic encounters, is an essential aspect of analytic treatments. Researchers, however, strove to abolish the role of particular clinicians in defining any disorder. They regarded the dominant analytic therapeutic model as the offspring of such discredited practices as mesmerism or hypnotism. In their view, optimal practice would replace clinical insight with reliable diagnostic criteria grounded in observable symptoms and decision rules that do not depend on the personal characteristics of either patients or clinicians. "The clinician's task is twofold: to determine the presence or absence of specific clinical phenomena, and then to apply the comprehensive rules provided for making the diagnosis," Spitzer declared.[27] This empiricist philosophy was incompatible with the foundational tenets of dynamic psychiatry, which relied on unobservable concepts (such as the unconscious), explanations that are unprovable, and treatments that are intuitive.

An additional split between clinicians and researchers involved the range of diagnoses the *DSM* should include. Clinicians seek to provide treatment for all who want it. They require a classification that encompasses the full range of problems they see in their practices, regardless of the evidence supporting the validity of any particular diagnosis. Psychodynamically oriented clinicians, who most commonly treated patients with neuroses or personality disorders, were especially concerned that a new *DSM* would limit the range of diagnoses and thus threaten their practices. They worried that a strict imposition of operational criteria "would not allow adequate diagnoses of the patients that analysts see" and would provide third-party payers an excuse to deny reimbursement.[28] Given that about half of the APA membership had psychoanalytic training at the time, the new manual could not ignore this large constituency. In contrast to clinicians, who strive for maximum diagnostic inclusivity, researchers are prone to use only clear-cut conditions with extensive empirical validation. They require well-defined, homogeneous groups before they can uncover the etiology, prognoses, and likely outcomes of any condition. Therefore, they tend to use a small number of categories and reject diagnoses that are not well established.

No diagnostic system can simultaneously realize inclusive clinical goals to maximize the number of treated individuals and exclusive research standards that limit diagnoses to unambiguous categories. Yet Spitzer had to develop a system that would serve the dual—and highly divergent—interests of researchers and clinicians. Unsurprisingly, the process leading to the *DSM-III* was highly contentious.

MODELS FOR THE *DSM-III*

The *DSM-II* was clearly an unsuitable foundation for a new diagnostic system based on principles of specificity. Instead, the *DSM-III* built upon two related prior efforts to develop explicit, empirically defined classifications. The first was the Feighner criteria, which originated in the Washington University Department of Psychiatry. In the 1960s this department was a rare outpost of medically minded psychiatric research. It was the major institutional base of the newly emerging research-oriented psychiatrists who defined themselves in opposition to the dominant analytic wing of the profession.

The Washington University group strove to use operational definitions to develop homogeneous categories of distinct mental illnesses. They advocated for a strict medical model that viewed each mental illness as consisting of clusters of observable symptoms that were separable from each other, from non-disordered conditions, and, especially, from idiosyncratic individual experiences. For them, anything worth studying had to be measurable; unless they could be quantified, subjective processes had no place in the diagnostic process. Diagnoses were not just important but were *the* central aspect of the discipline because all other psychiatric concerns—determining the etiology, prognoses, and treatment of patients—depended on the accuracy of its diagnostic system. In 1972, the Washington University psychiatrists published what came to be known as the "Feighner criteria" after the psychiatric resident who recorded the definitions of 14 different conditions for use among researchers.[29] Despite the fact that the extant research base for these criteria was minimal, this article became the most widely cited piece in the history of psychiatric journals.[30]

The second model for the new *DSM* was the research diagnostic criteria that Spitzer had developed in collaboration with the Washington University group and the NIMH. Like the Feighner criteria, the RDC was developed for use in psychiatric research, not in clinical practice. The RDC expanded the 14 Feighner diagnoses to 25 specifically defined conditions and many additional subtypes. For example, the RDC category of major depressive disorder contained 11 distinctive subclasses, such as minor depression with anxiety, which was a separate category from generalized anxiety disorder with significant depression. Spitzer was not shy in advocating for clear diagnoses. In the article that presented the RDC, he quotes Alvan Feinstein, a pioneer of psychiatric epidemiology: "In the field of diagnostic nosology, the establishment of operational criteria represents a breakthrough that is as obvious, necessary, fundamental and important as the corresponding breakthroughs in obstetrics and surgery when Semmelweiss, Oliver Wendell Holmes, and, later on Lord Lister, demanded that obstetricians and surgeons wash their hands before operating on the human body." Spitzer concluded the piece with enthusiasm, "The use of operational criteria for psychiatric diagnosis is an idea whose time has come!"[31]

Spitzer faced two major barriers in constructing a new *DSM* based on the Feighner and RDC models. One was that neither system was grounded in a strong evidence base; none of their diagnoses had consensual empirical support. The second was that the new manual had to use operational criteria that clinicians as well as researchers could use. Yet psychodynamic psychiatrists disdained the empiricist approach. For analysts, "psychiatric diagnoses made only on the basis of signs and symptoms, without a positive psychodynamically informed, coherent understanding of why the patient has developed the symptom at this time is second-rate diagnosis."[32] Spitzer's goal of developing a revolutionary new diagnostic manual faced formidable obstacles.

Designing the *DSM-III*

The immediate impetus to revise the *DSM-II* was identical to the stimulus behind the first revision. The World Health Organization had started the process of producing a new version of its diagnostic manual, the *ICD-9-CM*.[33] Because of his work as a consultant to the *DSM-II*, the APA Board of Trustees appointed Spitzer to lead the effort that would ensure the American manual conformed to the *ICD* revision. Spitzer had ambitious plans for the new manual. He realized that the competing aims of research-focused operational principles and clinical practice would not be easy to reconcile. The new manual would have to bridge a basic divide in the profession between researchers aligned with the Washington University tradition and the analysts: "It was, at base, a struggle over both the image and intellectual commitments of a profession seeking to fashion a paradigm for its discourse and work, a struggle over the relative status and authority of those working within distinct traditions," Spitzer wrote.[34]

The APA gave Spitzer vast control over the development of the *DSM-III*; he was entirely responsible for setting the agenda and organization of the revision. Prior *DSMs* emerged from a single undifferentiated committee consisting of about 30 members. In contrast, Spitzer created a far more elaborate structure for the *DSM-III*. He appointed all of the nearly 200 individuals involved in the revision. They were divided into a general task force of 19 members, 14 advisory committees concerned with particular classes of disorder, and various additional consultants. Spitzer was closely involved in the deliberations of all these committees and had to approve each of their proposals. He later recounted that he typed "every word" of the manual.[35]

Most of the DSM-III Task Force and advisory committee members were Spitzer's allies from Washington University, their students, and his colleagues at Columbia. They were far more attuned to research than clinical interests: "The people I appointed had all made a commitment to be guided by data," Spitzer reported.[36] None held divergent viewpoints; adherents to dynamic perspectives were not represented until the revision process was already well underway. As it turned out, task force and committee members often took harder-line positions against the analysts than did Spitzer, who was more prone to make concessions to them. Many were contemptuous of psychoanalysis and considered the *DSM-II* to be "an embarrassment to the profession."[37] Spitzer, however, was intensely aware that the new diagnostic manual would have to serve the entire psychiatric profession and compromises with analytically oriented clinicians were inevitable.[38]

Spitzer and his appointees were particularly worried about the varied and id-iosyncratic diagnostic processes that prevailed in psychiatric practice. All were committed to developing a manual based on explicit, descriptive, and empirical criteria that would ensure that psychiatrists spoke the same language, used reliable diagnoses, and employed conventional scientific methods. They believed that common rules were more likely than clinical intuition to lead to correct diagnoses and would have agreed with psychologist Paul Meehl's assessment that "the clinical interpreter is a costly middleman who might better be eliminated."[39] To realize this goal, they strove to develop "specific inclusion and exclusion criteria for each diagnosis. The clinician is required to use these criteria, regardless of his own personal concept of the disorder."[40]

As a consequence of this empiricist perspective, reliability—the extent to which different users agreed on the measurement of a specific diagnosis—became the driving force justifying the new manual. The developers of the *DSM-II* had paid little attention to this issue. Instead, they assumed that the variety of perspectives within psychiatry precluded general agreement on how to apply its diagnostic labels. For example, in regard to schizophrenia they lamented, "Even if it had tried, the Committee could not establish agreement about what this disorder is."[41] Unsurprisingly, studies found that the *DSM-I* and *II* diagnoses had poor reliability. For example, Aaron Beck's review indicated that just over half of psychiatrists agreed on a diagnosis when they examined the same patient. The result was that "psychiatric diagnosis at present is so unreliable as to merit very serious question when classifying, treating and studying patient behavior and outcome."[42] Another well-publicized study that presented a video of a socially awkward man to US and British psychiatrists found that the former were far more likely to diagnose schizophrenia in a case where the latter saw manic depression. By the 1970s, it appeared that what diagnosis a patient received was more due to idiosyncratic factors, cultural preferences, or random chance than to any rational classification system.[43]

In the mid-1970s Spitzer and other researchers filled psychiatric journals with calls to develop explicit and reliable diagnostic criteria. In their view, reliability was a prerequisite for validity: no theory about the actual nature of any mental disorder could be valid unless diagnosticians agreed on how to measure it. They argued that diagnoses that were not reliable could not possibly provide a foundation for psychiatry's ultimate goals to understand the etiologies, prognoses, and treatments for its various conditions. Adequate research required homogeneous groups of subjects, which in turn depended on reliable selection criteria. Yet Spitzer and others demonstrated that reliabilities for the major *DSM-II* categories were appallingly poor.[44] "Without reliability the system is completely random, and the diagnoses mean almost nothing—maybe worse than nothing, because they're falsely labelling. You're better off not having a diagnostic system," Allen Frances noted.[45]

Spitzer received a grant of about $100,000 from the NIMH to conduct field trials to test the reliability of the major diagnoses proposed for inclusion in the new manual. These trials, which employed a draft of the *DSM-III* to see whether pairs of volunteer psychiatrists could reliably use it, showed higher reliabilities than studies of *DSM-II* diagnoses and provided modest evidence of the usefulness of the revised system.[46] They played no role, however, in developing the new manual or in testing it against any alternative classification. "The field trials," Stuart Kirk and Herb Kutchins summarized, "were not trials in the sense that something was tested that might fail, but rather they were activities that would reinforce the solutions that already had been framed."[47]

By tying the justification of the new diagnostic system so closely to the achievement of reliability, Spitzer and his allies downplayed the fact that a reliable classification is not necessarily a valid one. Operational criteria can be completely reliable and simultaneously completely invalid. Medieval physicians, for example, might have all agreed on who they called witches. Perhaps most worrisome, the relatively small body of empirical research developed by the mid-1970s had not produced consistent findings that could lead to consensus on diagnostic criteria. "There was very little systematic research, and much of the research that existed was really a hodgepodge—scattered, inconsistent, and ambiguous. I think the majority of us recognized that the amount of good, solid science upon which we were making our decisions was pretty modest," task force member Theodore Millon observed.[48]

Although all the participants in the revision were committed to developing data-driven, empirical diagnoses, they did not agree on optimal criteria for even the few diagnoses with a substantial research base. Depression provides a good example.[49] A few researchers, most notably, British psychiatrist Aubrey Lewis, affirmed Kraepelin's contention that depression was a single disorder.[50] However, studies conducted with outpatient as well as inpatient samples rejected a unitary model of depression. Researchers from the 1950s through the mid-1970s generally agreed that psychotic (or endogenous) depressions constituted one distinct type of depression, but they held widely divergent opinions about the nature of nonpsychotic depressions. Some concluded that depression was binary, featuring a neurotic type as well as a psychotic one.[51] Others believed that three or more distinct types of neurotic depressions existed, although they differed on both the number and the nature of these states.[52] In addition, unlike the relatively homogenous signs of psychotic depressions, indicators of the neurotic types were heterogeneous and diffuse across studies. Depending on the study, neurotic depressions featured some combination of symptoms that reflected helplessness, low self-esteem, dysphoria, demoralization, anger, hostility, irritation, and disappointment reactions that resisted precise diagnostic schemes.[53]

Faced with this confusing situation yet needing to find some resolution, the advisory committees had no choice but to vote on criteria for each diagnosis. Donald Klein described the process to anthropologist James Davies:

> We had very little in the way of data, so we were forced to rely on clinical consensus, which, admittedly, is a very poor way to do things. But it was better than anything else we had. We thrashed it out, basically. We had a three-hour argument. There would be about twelve people sitting down at the table, usually there was a chairperson and there was somebody taking notes. And at the end of each meeting there would be a distribution of events. And at the next meeting some would agree with the inclusion, and the others would continue arguing. If people were still divided, the matter would be eventually decided by a vote . . . that is how it went.[54]

Given the absence of known pathogens or objective tests for any mental disorder, the *DSM-III* diagnoses emerged in the only possible way: through a messy procedure marked by arguments, heated discussions, and, eventually, reluctant consensus.

If the research base for psychiatry's core conditions was weak, it was virtually nonexistent for most proposed diagnoses. Yet the growing dependence of clinicians on third-party payment dictated that the manual's approach to diagnosis had to be inclusive, not exclusive. Spitzer realized that if the manual did not contain labels for the various kinds of problems practitioners were treating, there was no hope that the APA assembly would approve it. This meant that he had to defy the preferences of researchers for a more limited delineation of diagnoses. When the leader of the Washington University group, Samuel Guze, tried to implement the principle that the manual should contain only entities that follow-up studies had validated, he received the response: "If we do what you are proposing, which makes sense to us scientifically, we think that . . . we will give the insurance companies an excuse not to pay us."[55]

Spitzer adroitly found a good way to temper the conflict between the inclusionary views of clinicians and exclusionary tendencies of researchers. *All* of the manual's many conditions would follow the operational principles found in the Feighner criteria and RDC. The result was that the *DSM-III* embraced "as many conditions as are commonly seen by practicing clinicians," Millon wrote.[56] Spitzer, too, admitted, "If any group of clinicians had a diagnosis that they thought was very important, with a few exceptions, we would include it. That's the only way to make it acceptable to everyone."[57] The resulting manual contained 265 diagnoses, compared to 106 in *DSM-I* and 182 in *DSM-II*. Each was based on observable symptoms and self-reports rather than inferred theoretical processes, presumably minimizing the role of diagnostic subjectivity.

NEUROSIS

The controversy over the concept of neurosis was not as easily resolved. Spitzer and the task force had concluded that an essential aspect of the *DSM-III* would be theory neutrality. The importance of this principle was twofold. First, it provided the basis for rejecting the *DSM-I* and *II* approach, where analytic assumptions infused many core diagnoses. Second, it opened a pathway for the relatively new biological branch of psychiatry that might in time lead to important new etiological and treatment breakthroughs. Biological psychiatrists assumed that descriptions of each diagnosis without reference to their causes would eventually show how they "have specific genetic patterns, characteristic responses to drugs, and similar biological features."[58] This organic basis, however, would come through research that did not yet exist in 1980. A core principle of the new manual was thus that "the approach taken in DSM-III is atheoretical with regard to etiology or pathophysiological process."[59]

The issue of theory neutrality was at the heart of the single most contentious subject in the creation of the *DSM-III*: what to do about the status of the term "neurosis." This was the central analytic concept and was at the heart of both the *DSM-I* and *II*. Analysts did not view "neurosis" as a descriptive term but as the core process that explained why patients made poor adaptations to their environments. It implied the Freudian assumption that some problem arose because of unconscious conflicts remaining from childhood interactions. For analysts, removing neurosis from the manual would destroy the heritage of the past 80 years of psychiatric thinking. They justifiably believed that Spitzer and his task force were far more anti-analytic than theory neutral, warning that if the term was eliminated, there would be a "mass exodus from the DSM-III to the ICD-9-CM," which did have a section for neurotic disorders.[60] Prominent analyst Otto Kernberg baldly stated that the new manual would a "disaster. It is a straitjacket and a powerful weapon in the hands of people whose ideas are very clear, very publicly known, and the guns are pointed at us."[61]

Spitzer and his diagnostically inclined allies, however, viewed neurosis as intrinsically tied to a particular etiological formulation that violated the manual's intention to be theory neutral. Its inclusion in the *DSM-III* would signify that a condition stemmed from a particular set of psychodynamics. They also considered it a sloppy and heterogeneous word that could distinguish neither the neuroses from other types of mental disorder nor the different neuroses from each other.[62] Spitzer and the task force, therefore, were set on eliminating the term from the manual. Their determination was so strong that they ignored the fact that the purported purpose of the revision was to keep the *DSM* in line with the forthcoming changes in the *ICD-9*, which did contain a major category of neurotic disorders, which "are disorders without any demonstrable organic basis."[63]

The prolonged period of conflict between the analysts and the task force came to a head shortly before the *DSM-III* went before the APA Assembly for approval in 1979. The degree of anticipated opposition to abandoning the term "neurosis" was so strong that the assembly seemed likely to reject the new manual. Realizing the impasse could have disastrous consequences, Spitzer came up with an ingenious "neurotic peace treaty" to accommodate both sides.[64] "Neurosis" would remain in the *DSM-III* but would be used as a descriptive rather than an explanatory concept. Labels for the former psychoneuroses would read, for example, "Phobic disorders (or Phobic neuroses)" or "Obsessive compulsive disorder (or Obsessive compulsive neuroses)."

Many analysts were dissatisfied with this resolution. Two eloquently complained, "DSM-III gets rid of the castles of Neuroses and replaces it with a diagnostic Levittown."[65] Many task force members, too, were not pleased with the concession but ultimately accepted what Donald Klein called "a minor capitulation to psychoanalytic nostalgia."[66] The controversy also subjected the profession to considerable ridicule. Bayer and Spitzer observed how media accounts mocked the hullabaloo:

> Their accounts invariably portrayed psychiatric deliberations as peculiar, almost irresponsible. "Farewell to Neuroses: Mental Health Dictionary Drops Name," wrote the Detroit Free Press. In the Boston Globe, the tale was recounted in an ironic story entitled "Putting an End to Neurosis." Even the medical press could not resist the opportunity to make psychiatrists seem a bit silly. "No More Neuroses—Psychiatry Has Retired Them," was the headline in Medical World News. "Neuroses Banned" was the way in which Hospital Doctor told its readers of the events that had come to pass.[67]

Despite the derision, the compromise satisfied enough members of the assembly, which ratified the *DSM-III*.

Klein was correct when he called the maintenance of "neurosis" as a descriptive term a "minor capitulation." The *DSM-III* destroyed the former unified concept by dispersing the neuroses into separate affective, anxiety, dissociative, and somatoform (psychosomatic) classes. The demolition of Freudianism in psychiatry was so complete that the school of thought virtually disappeared among psychiatrists who trained after 1980.[68] Spitzer had won an intensely fought political battle. "The entire process of achieving a settlement seemed more appropriate to the encounter of political rivals than to the orderly pursuit of scientific knowledge. On each side of the controversy it was held that important scientific truths were at stake, and yet the situation had demanded, of those who found themselves in opposition, the adoption of strategic postures and the employment of the techniques of politics," he and Ronald Bayer concluded.[69]

DEFINITION OF MENTAL DISORDER

No prior diagnostic manual had contained any general statement about the nature of mental disorder itself. Classifications from the *Statistical Manual* through the *DSM-II* did not explain what the categories in the diagnostic system were categories *of*. Instead, they presented various types of mental illness without noting what qualities unified the particular conditions. Many DSM-III Task Force members saw no need for such a general statement. For example, Henry Pinsker observed that other medical specialties did not define "disease" itself.[70] Eminent British psychiatrist John Wing told Spitzer that a definition of "mental disorder" in the *DSM-III* was unnecessary because practitioners needed only descriptive definitions of each disorder to make adequate diagnoses.[71]

Spitzer's prior experiences, however, led him to believe that the *DSM-III* must contain some concept of mental disorder itself.[72] He was deeply concerned that the failure to include such a statement would leave the profession vulnerable to the widely heralded anti-psychiatric critique that the very notion of mental illness was a "myth."[73] Spitzer was also involved in a highly publicized controversy with psychologist David Rosenhan, who claimed that psychiatrists were unable to distinguish the mad from the sane. The biting title of Spitzer's critique of Rosenhan's article, "On Pseudoscience in Science, Logic in Remission, and Psychiatric Diagnosis," indicates the tenor of this debate. Yet, paradoxically, both Spitzer and Rosenhan had the same agenda: to show how the existing diagnostic system was thoroughly unreliable.[74]

A second source of Spitzer's insistence on developing a general definition of "mental disorder" was his immersion in the controversy over the *DSM-II*'s definition of homosexuality in the early 1970s discussed in the previous chapter. This imbroglio did not so much involve the particular criteria for this diagnosis as the question of what a mental illness itself is. Critics charged that there was no justification including homosexuality in a manual of psychiatric diagnosis at all. Advocates, who were mostly psychoanalysts, urged that "you cannot make pathology go away by removing its label."[75] Spitzer's active participation in the debates over anti-psychiatry critiques and whether homosexuality was a mental illness convinced him that, unlike its predecessors, the *DSM-III* must contain some definition of "mental disorder."

Spitzer stumbled in his initial attempt to construct this statement. His first proposal framed the issue around the claim that "mental disorders are a subset of medical disorders."[76] Although the *DSM* was the official manual for the psychiatric profession, other mental health professionals also had to rely on it for billing purposes. Spitzer's wording implied that psychiatrists were the only truly legitimate responders to mental disorders. Psychologists, in particular, expressed outrage over labeling mental disorders as "medical." Theodore Blau, the president of the American Psychological Association, wrote an angry letter to then-APA president Jack Weinberg, copying the presidents of the social work, nursing, and education associations.[77] Weinberg was unyielding in his response to Blau: "We believe it is inappropriate for you to attempt to tell us how we should conceptualize our area of professional responsibility."[78] Blau bluntly answered, "Candidly DSM-III, as we have seen it in its last draft, is more of a political position paper for the American Psychiatric Association than a scientifically-based classification system."[79] He put Weinberg on notice that the psychologists had decided to develop their own classification of what he called "behavioral disorders." The conflict escalated to the point that the psychologists' professional organization threatened to initiate legal action against the APA if the *DSM-III* included the claim that mental illnesses were medical disorders.

Spitzer eventually realized that the intense interprofessional enmity his definition created could threaten the credibility of the manual itself. He ingeniously altered it to state:

> In DSM-III each of the mental disorders is conceptualized as a clinically significant behavioral or psychological syndrome or pattern that occurs in an individual and that is typically associated with either a painful symptom (distress) or

impairment in one or more important areas of functioning (disability). In addition, there is an inference that there is a behavioral, psychological, or biological dysfunction, and that the disturbance is not only in the relationship between the individual and society. (When the disturbance is *limited* to a conflict between an individual and society, this may represent social deviance, which may or may not be commendable, but is not by itself a mental disorder.)[80]

The notion that a mental disorder was a "clinically significant behavioral or psychological syndrome or pattern" was acceptable to all the mental health professions, which dropped their opposition.

Spitzer's succinct definition both lays out what any mental disorder is—some clinically significant dysfunction that is accompanied by either distress or disability—and what it is not—conflicts between individuals and society or social deviance. Although none of its central concepts, like "clinical significance," "disability," "dysfunction," or "social deviance," is well defined, the statement provided a useful general standard for what a mental disorder itself is. Indeed, Spitzer considered the vagueness as a benefit: "This definition has the advantage of not being precise. There's no pretense that it makes a sharp line between disorder and no disorder."[81] Spitzer developed this statement, however, independently of the advisory committees that constructed the criteria sets for the various disorders. Therefore, many of the particular *DSM* diagnoses ignored it. While Spitzer's definition provided notable fodder for philosophers of mental disorder for decades to come, it had little influence on the actual content of the manual.[82]

The Outcome

The process leading to the *DSM-III* provided the venue for psychiatric researchers to wrest control of the profession from the previously dominant analytic group. The manual's professional achievement was to mold a small, inconsistent, and highly contested body of findings into a scientific-looking and precise group of mental disorders. *DSM-III* seemed to be an effective retort to critics who charged that the field was reliant on subjective, idiosyncratic preferences more than objective, evidence-based knowledge. It provided psychiatry with the specific diagnoses it required before it could be accepted as a legitimate medical specialty. Spitzer and his task force's work was so effective that it took two decades before researchers realized that the manual they had constructed was built on a remarkably shaky foundation.

FOUR | The *DSM-III*

The third version of the *DSM* thoroughly contrasted with the previous two manuals. The *DSM-I* and *II* were small, spiral-bound booklets of fewer than 150 pages. Neither needed even 40 pages to present all of its brief, nonspecific diagnostic definitions. In contrast, the *DSM-III* was an imposing hardback volume, running to nearly 500 pages. It expanded the space devoted to particular diagnostic categories nearly tenfold, to 304 pages. Detailed and clear-cut symptom lists as well as decision rules that yielded standardized diagnoses accompanied most of these entities. "In style and appearance," sociologist Owen Whooley observes, "DSM-III suggested a confident, well-informed scientific grasp of mental illness, a physical embodiment of psychiatry in the scientific image."[1]

An Empiricist Manual

Visible symptoms played a sharply divergent role in the *DSM-III* diagnoses compared to those of its predecessors. The dynamic psychiatrists who developed the *DSM-I* and remained influential in the *DSM-II* scorned the use of overt symptoms to make a diagnosis because they were more likely to hide than to reveal the nature of any disorder. The *DSM-III*'s basic transformation was to develop a model that equated observable and measurable indicators with the presence of a mental disorder. It provided operational definitions that specified what combinations of symptoms were necessary to make each of its nearly 300 diagnoses. Such clear-cut criteria sets presumably ensured the reliability of each condition. In 1984, leading researcher Gerald Klerman announced, "The reliability problem has been solved."[2]

The foundational role that the *DSM-III* gave to external symptoms has led it to be seen as a "neo-Kraepelinian" manual.[3] Spitzer himself denied being a neo-Kraepelinian, although many members of his task force did embrace this label.[4] Yet the *DSM-III*'s approach diverged significantly from Kraepelin's. Kraepelin vigorously argued *against* using symptoms that are apparent at any particular time as a basis for categorization. He was interested in the dynamic unfolding of symptoms, not in their static manifestations. As the example of syphilis indicated to him, symptoms changed over time. The core of the diagnostic process was to understand how the same underlying disorder was present although markedly different symptoms appeared at different stages of the disease. In addition, Kraepelin firmly believed that true mental disorders were grounded in biology, specifically, in heredity or toxins. Finally, he admitted only well-validated conditions into the psychiatric canon and emphasized just two: dementia praecox (schizophrenia) and manic depression.

In contrast, the *DSM-III* criteria sets provided static definitions that rarely took into account the dynamic course of an illness. In addition, they did not assume that mental disorders had a biological, or any other, foundation; the manual was theory neutral and compatible with all causal perspectives. Finally, unlike Kraepelin's restricted focus on a small number of distinct entities, the *DSM-III* contained nearly 300 separate diagnoses, many of which overlapped. Aside from the assumption that accurate diagnoses must begin with observing visible symptoms, the *DSM-III* bore little resemblance to Kraepelin's model.

The most accurate overall description of the manual seems to be "empiricist." Philosopher of science Carl Hempel developed perhaps the most influential statement of this perspective. In his view, objective criteria are the first, foundational stage of the scientific enterprise. The starting point of empiricism in medicine is the presence of clearly defined diagnostic categories. The purpose of observation is to eventually develop theoretical explanations, predictions, and expected outcomes of what is being observed.[5]

The vastly greater role of observable symptoms and specific decision rules in the *DSM-III* pointed to a second difference from earlier diagnostic manuals. Its criteria sets almost always focused on patients' *current* presentation of their problems. This tacitly distinguished the manual from the prior two *DSMs*, which gave far more importance to the way psychodynamics from past experiences produced symptoms. It also differed from Kraepelin's focus on the future course and outcome of any present constellation of symptoms. The *DSM-III* was far more rooted than its predecessors in the here and now rather than in either the historical origins or the consequent unfolding of symptoms.

Another key feature of the *DSM-III* was to separate rather than join diagnostic categories. The *DSM-I* and *II* strove to funnel diagnoses into a hierarchy where individuals had a single, dominant condition. In contrast, the *DSM-III* usually assumed that people who met the symptom criteria for more than one diagnosis had multiple, separate conditions. There was no need to consider what the primary diagnosis was. Although the manual's introduction states, "In DSM-III there is no assumption that each mental disorder is a discrete entity with sharp boundaries (discontinuity) between it and other mental disorders," its very essence was to try to provide the types of descriptions that created sharp boundaries with other conditions. Despite this disclaimer, the manual framed its diagnoses as discrete, discontinuous entities.[6]

The *DSM-III* also developed a "multiaxial" system. In addition to the bulk of conditions that it placed on Axis I, it contained a separate section, which it called "Axis II," for 11 personality and developmental disorders (see below). It also contained three additional axes for physical disorders (III), psychosocial stressors (IV), and level of patient functioning (V). Although the task force touted the multiaxial system as an innovative and more comprehensive way to understand mental disorders, in fact, it resurrected a similar system that appeared in William Menninger's *Medical 203*. In any event, Axes III, IV, and V turned out to be largely superfluous; neither clinicians nor researchers made much use of them.

In addition to its empiricist, cross-sectional, categorical, and multiaxial approach, the new manual was notable for what it did *not* do. Despite the greater number of specific diagnoses in the *DSM-III*, the new manual rarely expanded the realm of mental disorder itself.[7] It seldom pathologized conditions that had been outside of the range of the much broader *DSM-I* and *II* descriptions. For example, neither of the initial *DSMs* contained a diagnosis of social phobia. It first appeared in the *DSM-III*, which defined it as "a persistent, irrational fear of, and compelling desire to avoid, a situation in which the individual is exposed to possible scrutiny by others and fears that he or she may act in a way that will be humiliating or embarrassing."[8] Yet someone who met these criteria could have received *DSM-II* diagnoses such as "phobic neurosis," "schizoid personality," or "inadequate personality." The *DSM-III* significantly increased the number of distinctly defined disorders, but it would have been hard for it to broaden the scope of disordered conditions that the earlier manuals already contained. The *DSM-I* and *II* were more than able to incorporate the wide range of problems that patients presented in the 1950s, 1960s, and 1970s.

Major Depressive Disorder

The most impactful change for any particular *DSM-III* diagnosis regarded depression. For the entire course of psychiatric history before 1980, depressive disorder was divided into two distinct general types.[9] The first, melancholic or endogenous depression, was an exceptionally severe but seldom-found condition marked by extended periods of extreme low mood, nonresponsiveness to external events, feelings of worthlessness and guilt, and slowness of thought and movement that could deteriorate into psychotic periods marked by delusions and hallucinations. Many observers considered this form of depression to have a genetic or biochemical basis. The condition rarely responded to psychotherapy but could improve after somatic treatments including electroconvulsive therapy (ECT) and antidepressant drugs. However, in contrast to the tranquilizers, the antidepressant MAOIs and imipramine garnered little public attention because the severe conditions they targeted were uncommon.[10] The *DSM-I* and *II* placed this condition in the affective psychotic category, making clear that it was distinct from depressive neurosis.[11]

The second type of depression had a variety of names, including "reactive depression," "neurotic depression," or simply "depression." Since the late nineteenth century, it was attached to the neurasthenic tradition that American neurologist George Beard first elaborated. In contrast to the disease-like qualities of melancholy, the neurasthenic package of symptoms was a heterogeneous and diffuse mix of fatigue, anxiety, worries, and various bodily aches and pains.[12] These often arose after some psychosocial stressor and lacked the extreme seriousness, psychotic features, and nonresponsiveness to external events that marked melancholy.[13] Moreover, neurotic depressions were typically accompanied by anxiety, tension, and general unhappiness that were often transient. Unlike melancholic patients, who were commonly hospitalized, general physicians or outpatient psychiatrists typically treated reactive depressions. Psychotherapy, frequently accompanied by a minor tranquilizer, was the most common response to them.

The first two *DSMs* clearly divided neurotic depression, which they considered one type of response to underlying anxiety, from the affective psychoses. Because anxiety, not depression, was the chief characteristic of all the neuroses, neurotic depression was rarely diagnosed: in 1962 the National Disease and Therapeutic Index reported that it accounted for fewer than 1 in 10 diagnosed psychoneuroses.[14] Instead, as the previous chapter indicated, anxiety was far and away the most frequent diagnosis during the 1950s and early 1960s. Sociologist Charles Kadushin's findings about why people went to outpatient psychiatrists in 1959–60 captured the approach to labeling in this era: "If this is not the age of anxiety it is certainly the age of the recognition and treatment of anxiety through psychiatry."[15]

In the early 1960s, however, drug companies began to target depression as well as anxiety. In a signal event in 1961, Merck, the maker of the antidepressant Elavil, distributed psychiatrist Frank Ayd's *Recognizing the Depressed Patient* free of charge to every physician in the United States.[16] Rates of diagnosed depression rose rapidly in the 1960s, doubling from just 4 million in 1962 to 8 million in 1968. By the early 1970s, depression reached parity in diagnosis with anxiety and by the mid-1970s had surpassed it as the most commonly diagnosed mental illness.[17]

This upsurge of interest in depression corresponded to a growing split between psychiatric researchers and clinicians. In the 1950s and 1960s, biologically oriented researchers coalesced around the study of depression as a way to differentiate themselves from those with an analytic focus on anxiety. Their theories emphasized neurochemistry and the brain, not early childhood and life experiences.[18] Moreover, their views were grounded in specificity; they strove to find the particular neurotransmitters (e.g., dopamine, noradrenaline, serotonin) associated with depression, a perspective that thoroughly contrasted with the capacious role that anxiety played for analysts. The tricyclics and MAOIs were promoted as particular remedies for depression, not for any other disorder: "These new drugs usually benefit a *specific* mental illness. In general, antipsychotic drugs (e.g., Thorazine) help schizophrenics, but not depressives; antidepressant drugs help only depressives."[19]

The *DSM-II* split affective psychosis from depressive neurosis on the basis of whether the onset of the disordered mood is "related directly to a precipitating life experience."[20] The DSM-III Task Force did not have a unified position regarding this division. Donald Klein, a central member, observed that the presence or absence of a triggering life event did not accurately characterize the difference between the two depressions because situational circumstances could provoke melancholic as well as neurotic depressions. For him, the key issue was not whether some environmental stressor precipitated the condition but instead whether the depression became self-perpetuating once it arose. Therefore, Klein maintained that the crucial distinction was, *after* emerging, how the condition responded to changing environmental conditions: "Once the episode is underway, it is autonomous, that is unresponsive to changes in the initiating circumstances. If the patient with a depressive episode [as opposed to a depressive reaction] regains his job the illness continues."[21] Therefore, Klein urged the continuation of the two depressions tradition but without the accompanying emphasis on a precipitating life event as the distinguishing factor.

A number of other task force members also rejected the endogenous/exogenous distinction. In contrast to Klein, however, they argued that there was only a single form of depression, marked by its presenting symptoms. Spitzer made an un-Kraepelinian decision and sided with this group against Klein: "The Cross sectional symptomatic picture of Major (full syndrome) [Depression] takes precedence over the course."[22] What counted was the current presentation—not the history or future course—of the depressive condition.

The *DSM-III* was the first diagnostic manual to have a separate category for affective disorders. Major depressive disorder (MDD) was at the heart of this new category. MDD combined in a single entity what had been separate psychotic and neurotic forms of depression.[23] An MDD diagnosis required the presence for at least a two-week period of dysphoric mood or loss of interest or pleasure in usual activities and four or more additional symptoms out of the following eight: (1) weight gain or loss or change in appetite, (2) insomnia or hypersomnia (excessive sleep), (3) psychomotor agitation or retardation (slowing down), (4) decreased sexual drive, (5) fatigue or loss of energy, (6) feelings of worthlessness or excessive or inappropriate guilt, (7) diminished ability to think or concentrate or indecisiveness, and (8) recurrent thoughts of death or suicidal ideation or suicide attempts.[24] Yet an MDD diagnosis did not necessitate the presence of any of the final three more serious symptoms.

The MDD diagnosis was not based on well-established empirical findings. In fact, it ignored previous research on the essential qualities of depressive disorders: "A depression is judged to be pathological if there is insufficient specific cause for it in the patient's immediate past, if it lasts too long, or if its symptoms are too severe," one summary read.[25] In contrast, the *DSM-III* criteria ignored the context in which symptoms arose, required just a short duration, and did not necessitate any severe symptoms. The MDD criteria did recognize the importance of context in a single instance. It excluded from diagnosis symptoms that weren't unduly severe or prolonged following an episode of bereavement.[26] Bereavement, however, was the sole exception: symptoms that resulted from any other kind of stressor could meet the MDD standard.

Bereavement aside, what accounts for MDD's otherwise purely acontextual symptom criteria? The affective disorders advisory committee took the principle that the *DSM-III* should be theory neutral to mean that it should purge all etiological assumptions from the MDD diagnosis. However, the committee went overboard and mistakenly assumed that terms such as "excessive," which had been part of the *DSM-II* criteria for neurotic depression, were also etiological and so purged them as well. This does not seem to have been an intentional decision or even one that the committee explicitly considered in its deliberations. Instead, it was an inadvertent result of efforts to improve the reliability of diagnoses through excising criteria that could not be precisely operationalized and were, therefore, subject to idiosyncratic interpretations.[27]

Because virtually all previous research indicated that melancholic or psychotic depression was distinct from other forms of depressive disorder, the MDD criteria did include a subclass of "with melancholia."[28] This category, however, was not a separate disorder but a type of MDD: melancholics and former neurotics alike had to meet the five symptoms, two-week duration requirement for MDD. Moreover, the melancholic subtype did not require any of the psychotic features that had marked previous characterizations of this condition. In effect, the *DSM*'s diagnostic criteria for MDD abandoned the separation of melancholic and reactive depressions that had persisted for centuries. Instead, it combined them into a single, extraordinarily heterogeneous diagnosis. Someone who had been severely depressed for years, could not leave her bed, and had continuous thoughts of worthlessness had MDD, as did an adolescent who felt depressed and unable to feel pleasure, had trouble sleeping, and lost his appetite and concentration after his girlfriend broke up with him two weeks before.

The *DSM-III*'s solution to the excised category of neurotic depression was also unprecedented. Spitzer had originally wanted to pair a type of "minor depression" with the "major depression" of MDD but faced opposition from psychiatrists who worried that any condition bearing the title "minor" would not be reimbursable.[29] Instead, Spitzer invented a new category called "dysthymia (or neurotic depression)" that was supposed to replace the *DSM-II*'s psychoneurotic depression.[30] Dysthymia, however, was unique in psychiatric history. Its symptom criteria, which required 3 of 13 possible indicators, were relatively easy to meet. For example, someone with insomnia, low energy, and low self-esteem could receive the diagnosis. However, the symptoms had to persist for two *years*. Dysthymic symptoms fit the former reactive category, but their necessary duration was better suited to the former psychotic condition. The prolonged suffering that characterized dysthymia coupled with its characteristics, such as brooding, pessimism, inadequacy, and chronic tiredness, meant that it much more closely resembled a personality condition than an affective disorder.[31] The result was that any depressive condition that lasted for just two weeks but less than two years would be called major depressive disorder.

In short, the *DSM-III* diagnostic criteria for depression were a mess. MDD combined into a single entity two depressions, one of which was widespread but often not severe and the other incapacitating but uncommon. Transient and ubiquitous reactive depressions were diagnostically equivalent to long-standing and rare melancholic ones. "There are two kinds of depression," Edward Shorter explains, "as different as tuberculosis and mumps: it makes no sense to lump both of them together under the general term 'depression.'"[32]

Serendipitously, MDD's combination of melancholic with reactive depression turned out to be the *DSM-III*'s greatest gift to psychiatry. Population surveys, which did not distinguish the two, unsurprisingly found that huge proportions of people met the MDD criteria. The major survey of mental disorder in the United States conducted after 1980 found that over 20 percent of community members had suffered from MDD.[33] A Janus-faced disorder, MDD allowed researchers to downplay its severity when explaining how it could afflict such a substantial portion of the population. For this purpose, it was the "common cold" of psychiatry.[34] Yet, when it was advantageous to emphasize its devastation, depression could be seen as "the major scourge of mankind."[35] The World Health Organization declared depression to be the world's most disabling condition after it combined the large group of people who met the MDD criteria and assumed the severity was comparable to paraplegia or blindness![36] While this might be justified for the relatively small number of cases of melancholic depression, the same can hardly be said for someone who was sad, fatigued, unable to concentrate, and had sleep and appetite problems for just two weeks after facing a loss event.

The result of the MDD criteria was that "depression" largely replaced what past eras had called "anxiety," "nervous tension," or "stress" as the most common label applied to the psychic consequences of problems of living. For patients it served to both give them a framework to understand their problems and secure them medications that would come to be called "antidepressants." For psychiatrists and other mental health professionals, it was a reputed "major" disease that justified a huge market for their services. Governmental attempts to destigmatize mental illness trumpeted the huge estimates of how many people they affect.[37] Advocates could also argue that, because the vast majority of people with "depression" did not receive professional treatment, there was a huge "unmet need" for mental health services. Perhaps most important, the expansive range of MDD made it the point of entry for the next spectacularly successful class of psychotropic drugs, the selective serotonin reuptake inhibitors (SSRIs). After 1980, "depression" would displace "anxiety" as the moniker for the vast realm of personal problems and life difficulties linked to the neurasthenic tradition but with the hefty label of "major depressive disorder." This resulted from not only the broad definition of MDD but also the simultaneous carving up of the formerly dominant anxiety category.

ANXIETY

The *DSM-III* diagnostic criteria for depression become even more curious when compared to those for the anxiety disorders. Kraepelin believed that anxiety was so ubiquitous and appeared in such a broad range of other mental and physical disorders that it did not warrant consideration as a distinct condition. Unlike his German rival, Freud placed anxiety at the center of the neuroses. The *DSM I* and *II* embodied the Freudian perspective that anxiety was a consuming force that took on a wide range of manifestations. During the 1950s and 1960s, it was psychiatry's most commonly diagnosed condition. Indeed, the postwar period itself was famously called "the age of anxiety."[38]

The very public political battle between the analysts and the task force focused on the *DSM-III*'s attempts to eliminate the term "neurosis" from the manual. Behind the scenes, Spitzer and the task force engaged in a less visible slaughter of "anxiety." One reason for the downfall of anxiety in the *DSM-III* stemmed from Spitzer's creation of separate advisory committees for the anxiety and the affective disorders, ensuring their criteria would not overlap. Another reason stemmed from the anxiety committee's goal of destroying the pervasive analytic concept of this process.[39]

The *DSM-I* and *II* emphasized how all anxious states, indeed all neuroses including depressive ones, were different manifestations of a single underlying force. In contrast to its consolidation of distinct depressive conditions within the MDD diagnosis, the *DSM-III* carved anxiety into nine different conditions: agoraphobia with or without panic attacks, social phobia, simple phobia, panic disorder, generalized anxiety disorder, obsessive-compulsive disorder, posttraumatic stress disorder, and atypical anxiety disorder.[40] Moreover, there was no central anxiety condition. Generalized anxiety disorder (GAD), which was "generalized, persistent anxiety of at least 1 month's duration," might have played this role. However, the *DSM-III* defined GAD as a residual category, not to be diagnosed when any other mental disorder was present.[41] Because it seldom arose in the absence of some other anxiety or other mental disorder, the manual's decision rules ensured that GAD diagnoses would rarely occur. It was "an atavistic ghost of its predecessor that can hardly stand alone as a diagnostic entity," anxiety expert Peter Tyrer concluded.[42]

Paradoxically, a number of these discrete anxiety conditions echoed the careful delineations Freud made in the 1890s, before he developed psychoanalysis.[43] In that decade, he described the specific psychic components of threat, irritability, and inability to concentrate and the somatic components of heart palpitations, breathing problems, tremors, sweating, and gastrointestinal disturbances that remain the central components of anxiety disorders. For example, Freud's description of anxiety attacks contained 10 of the 12 symptoms listed in the *DSM-III* diagnostic criteria for panic disorders. Likewise, Freud's description of anxious expectation, a free-floating state of nervousness and apprehensiveness, closely resembles the GAD diagnosis that emerged in the *DSM-III*:

> A woman who suffers from anxious expectation will imagine every time her husband coughs, when he has a cold, that he is going to have influenzal pneumonia, and will at once see his funeral in her mind's eye. If when she is coming towards the house she sees two people standing by her front door, she cannot avoid the thought that one of her children has fallen out the window; if the bell rings, then someone is bringing news of a death, and so on.[44]

The severance of the anxiety conditions from each other coupled with the unification of depression within the MDD diagnosis guaranteed that depression would replace anxiety both as the heir to the neurasthenic tradition and as psychiatry's central diagnosis in the post-*DSM-III* era.

A second consequence of its strict division of anxiety from depressive conditions was that the *DSM-III* abandoned what is probably the most common presentation of psychic distress—mixed states of anxiety and depression.[45] The *DSM-I* and *II* had made depressive neuroses a subset of anxiety. Likewise, the *ICD-9* diagnosis of neurotic depression stated, "Anxiety is frequently present and mixed states of anxiety and depression should be included here."[46] The *DSM-III*, however, regarded someone who met the respective criteria for both depression and an anxiety disorder as having a "comorbid" diagnosis, where two distinct conditions were present. Someone who failed to meet the full criteria for each condition would not be diagnosable at all. The absence of a mixed anxiety/depression diagnosis was a rare case where the *DSM-III* was overly exclusive instead of inclusive.

Psychoses

Since the earliest nineteenth-century systems, the psychoses were at the heart of psychiatric classifications. Diagnosticians did not view them as simply more severe types of mental disorder but as different in kind than other conditions. People with psychoses were disconnected from social reality; their bizarre and sometimes violent behaviors made them the bulk of inpatient populations. Yet the psychoses presented in a variety of ways; thus, intense debates marked their division from each other and from other conditions.

Spitzer was intensely concerned with psychotic diagnoses. One reason was his goal to counter the anti-psychiatry movement. In the late 1960s and early 1970s the psychoses, schizophrenia in particular, sprang to prominence as the major target of anti-psychiatric critiques. R. D. Laing, for one, questioned whether schizophrenia was an illness at all and celebrated the often-profound insights it could lead to. Rosenhan's critique, too, focused on psychiatry's inability to define the condition.[47]

Spitzer's second aim was to reduce the volume of schizophrenia diagnoses, which dominated the labeling of psychoses in American psychiatry at the time.[48] One factor contributing to the deterioration of psychiatry's prestige in the 1970s was the US-UK cross-national studies about the reliability of diagnoses. This well-designed project found striking results: for example, about 62 percent of US psychiatrists compared to just 34 percent of British psychiatrists diagnosed patients with schizophrenia; the comparable figures for depressive psychoses were 5 and 24 percent.[49] In addition, Spitzer was influenced by the Washington University group's conviction that electroconvulsive therapy could successfully treat severe depressions. This led them to prioritize diagnoses of depression over those of schizophrenia in cases where both affective and psychotic symptoms were present.[50] In 1978 Jean Endicott and Spitzer published the Schedule for Affective Disorders and Schizophrenia (SADS), which aimed to improve the accuracy of these diagnoses.[51]

The *DSM-II* had separated psychoses associated with organic brain syndromes from psychoses not attributed to physical conditions.[52] The latter category contained schizophrenia, major affective disorders, paranoid states, and other psychoses. The manual devoted three pages to general descriptions of 11 types and four subtypes of schizophrenia, all of which were "disturbances of thinking, mood and behavior . . . misinterpretations of reality and sometimes to delusions and hallucinations."[53] It separated affective disorders into involutional melancholia that arose during or after middle age, manic depression (with five subtypes), and other major affective disorder. Finally, paranoid states (with three subtypes) were formed around a single delusion.

In contrast, the *DSM-III* did not make a basic distinction between psychotic and other conditions. It severed the connection that previous manuals made that put schizophrenia, manic depression, and paranoia in the same psychotic class. Instead, it placed schizophrenic, affective, and paranoid disorders along with a class of psychotic disorders not elsewhere classified in separate categories. The psychoses thus constituted 4 of the manual's 16 classes of mental disorders. The 23 pages the manual devoted to the schizophrenic disorders provided highly detailed descriptions of the common signs of this category. To address the concern that American psychiatrists overdiagnosed schizophrenia, Spitzer limited the diagnosis to persons who showed signs of the illness for a continuous period of at least six months.[54] Notably, in contradistinction to the manual's major thrust, it reduced the number of subtypes from the *DSM-II*'s 11 to just 5 (disorganized, catatonic, paranoid, undifferentiated, and residual). The stricter criteria for schizophrenia in the *DSM-III*, coupled with the manual's loose definition of MDD, led diagnoses of schizophrenia to decline and those of affective disorders to increase after 1980.[55] "One of the more irritating consequences of DSM-III," Donald Klein bemoaned, "has been the plague of affective disorders that have descended on us."[56]

The affective disorder category was an amalgam of psychotic bipolar conditions, the extremely heterogeneous MDD diagnosis, and the "minor" condition of dysthymia. It was segregated from schizophrenia, creating the difficulty of what to do with states that shared symptoms of thought and mood disorders.[57] The *DSM-I* and *II* had placed the combined schizoaffective type under the schizophrenic category, but many psychiatrists thought it more closely resembled an affective condition. For example, the Feighner criteria disallowed people who met criteria for an affective disorder from receiving a diagnosis of schizophrenia.

Yet evidence was accumulating that the relationship between schizophrenia and manic depression was far more complex than Kraepelin's renowned division of the two on the basis of their sharply divergent courses had initially allowed. There was often considerable overlap not just in the symptoms of thought and mood disorders but also in their trajectories. Some psychiatrists complained that the separation of the psychoses into schizophrenic disorders of thought and affective disorders of mood was "totally arbitrary."[58] In the end, Spitzer and the advisory committee simply threw up their hands and decided it was impossible to define schizoaffective disorder, which is the only *DSM-III* condition without any diagnostic criteria.[59]

Personality Disorders

Just as much as the neuroses, the personality disorders embodied the clash between clinicians and researchers. Aside from the psychoneuroses, they were the most commonly diagnosed conditions among outpatients in the *DSM-I* and *II* era. By 1970 this group accounted for a full quarter of psychiatric diagnoses in outpatient settings.[60] Personality disorders presented potent diagnostic challenges. They were marked not so much by particular symptoms as by deeply embedded character traits that were indistinguishable from who a person *is*. These essential qualities of individuals were difficult to change and were often unresponsive to traditional therapies. Yet the diagnosis of personality disorders was vital to clinicians because the conditions affected so many of their clients, especially those who stayed in therapy for long periods of time. They were so central to dynamic theory and practice that the *DSM-II* contained 11 different types.

Unlike clinicians, many psychiatric researchers disdained the personality disorders. They viewed them as unmeasurable, nonspecific, and value laden, and they did not consider them true mental disorders. Prominent psychiatrist Aaron Beck contended that the notion of "personality disorder" itself was a "construct so artificial and removed from observables, that it is probably of little utility and, even worse, it is probably a misleading fiction."[61] Many of these disorders seemed to be similar to mood, anxiety, or psychotic conditions rather than constituting independent entities. In addition, they were often nonresponsive to drug treatments and so were of little interest to biologically oriented psychiatrists and pharmaceutical companies. Others considered the personality disorders to be predisposing factors to other mental disorders, but not disorders in themselves.[62] Some worried that, although the *DSM*'s general definition of mental disorder required the presence of a "dysfunction," it was not clear what a dysfunction of personality was a dysfunction *of*. Instead, many of these conditions such as "antisocial personality disorder" seemed more related to "conflicts between an individual and society," which the definition explicitly excluded.

Despite their resistance to definition through operational criteria, the group of personality disorders was so important to clinicians that Spitzer had no choice but to somehow retain them in the *DSM-III*. As a sympathetic psychiatrist noted, removing this category from the manual "would be a grade 4 to 5 psychosocial stressor for many of our brethren who are still in a state of mourning over the loss of the 'neuroses.'"[63] Spitzer rose to the occasion and constructed an original solution to the competing perspectives of clinicians and researchers. The personality disorders would remain in the manual, but on an "Axis II," separate from other mental disorders that were found on "Axis I."[64] This resolution allowed researchers to keep their distance from these conditions while at the same time providing clinicians with the opportunity to treat and bill for them.

Aside from their placement on a separate axis, the *DSM-III* diagnoses of the personality disorders did not feature the radical reconfigurations found in the affective and anxiety classes. As with all its classifications, the *DSM-III* criteria for its 12 personality disorders were more specific and elaborate than earlier definitions. The new manual also recategorized a few of the *DSM-II* disorders, dividing inadequate personality into separate avoidant and dependent types; eliminating the hysterical, explosive, and asthenic (lack of energy) classes; and adding schizotypal, histrionic, narcissistic, and borderline categories. Although these were major changes, they did not essentially modify how previous *DSMs* defined the personality disorders.

The *DSM-III* introduced the personality disorders section with the caution that "it is only when *personality traits* are inflexible and maladaptive and cause either significant impairment in social or occupational functioning or subjective distress that they constitute *Personality Disorders*."[65] It could not, however, specify why any of the personality disorders were "behavioral, psychological, or biological dysfunctions" as opposed to "social deviance" or "conflicts between an individual and society."

All the particular criteria sets for the personality disorders were riddled with negative value judgments. For example, paranoids were guarded and secretive, avoided accepting blame, questioned the loyalty of others, and lacked a sense of humor. Schizoids were emotionally cold and aloof and were indifferent to the feelings of others. Histrionics, in contrast, were overly dramatic, craved excitement, and constantly drew attention to themselves. Those with antisocial personalities virtually embodied conflicts between individuals and society: they had long histories of "lying, stealing, fighting, truancy, and resisting authority" that began in childhood. Other diagnoses, such as compulsive personality disorder, were also defined through their conflicts with those around them: "They stubbornly insist that people conform to their way of doing things." Likewise, passive aggressive people displayed "resistance to demands for adequate performance in both occupational and social functioning." The *DSM-III* criteria left no doubt that those with personality disorders were unpleasant, difficult, or dangerous people; it was unclear, however, why they had "mental disorders" as the manual defined the term.[66]

The new category of borderline personality disorder was especially hard to pin down: "The essential feature is a Personality Disorder in which there is instability in a variety of areas, including interpersonal behavior, mood, and self-image. No single factor is invariably present. Interpersonal relations are often intense and unstable, with marked shifts of attitude over time."[67] Theodore Millon called it a "bad and misleading label."[68] Moreover, many saw the borderline syndrome not so much as a distinct character type as a feature that cut across all of the personality disorders wherein behavior was more severe than neurosis but less so than psychosis.[69] Thus, they were on the "borderline." In this sense, borderline was an indicator of severity, not of a particular character type. Donald Goodwin, who wrote the leading psychiatric textbook in the Washington University tradition, remarked, "The borderline syndrome is a mess. . . . In short, in my opinion, the borderline syndrome stands for everything that is wrong with psychiatry [and] the category should be eliminated."[70] Far from being eliminated, the borderline diagnosis gained cultural prominence through Susanna Kaysen's best-selling memoir, *Girl Interrupted* (1993). Interestingly, Kaysen learned of her condition while reading the *DSM-III* in her local bookstore.

Although both clinicians and researchers accepted Spitzer's solution to place the personality disorders on a separate axis from other mental disorders, these conditions were opposed by a third group: feminists, who were deeply suspicious of what they regarded as psychiatry's pathologization of women. Although manifestoes such as Betty Friedan's *The Feminine Mystique* (1963), Kate Millett's *Sexual Politics* (1969), and Germaine Greer's *The Female Eunuch* (1970) attacked psychiatry's oppressive control over women during the *DSM-I* and *II* period, these manuals were so invisible that feminist critiques did not target them directly. The *DSM-III* process, however, antagonized many feminists.

Several of the personality disorders were shot through with negative value judgments of stereotypical feminine qualities. One was histrionic personality disorder, which was "diagnosed far more frequently in females than in males." Among its criteria were terms such as "self-dramatization," "vain and demanding," "dependent, helpless, constantly seeking reassurance."[71] Dependent personality disorder was another: "The essential feature is a Personality Disorder in which the individual passively allows others to assume responsibility for major areas of his or her life because of a lack of self-confidence and an inability to function independently; the individual subordinates his or her own needs to those others on whom he or she is dependent in order to avoid any possibility of having to be self-reliant."[72] Such diagnoses seemed to embody adverse value appraisals of stereotypical gender roles more than symptoms of a mental disorder. Feminist opposition to these conditions would escalate in the *DSM-III* revision process that the next chapter discusses. The other types of personality disorders were less conventionally feminine but equally prone to negative value judgments.

Despite their many diagnostic flaws, the personality disorders were simply too important for clinicians to eliminate them from a classification of mental disorders. The issue of what to do about homosexuality in the *DSM-III* raised many of the same issues but resulted in an entirely different outcome.

Psychosexual Disorders

As chapter 2 discussed, the question of whether homosexuality should be classified as a mental disorder had propelled the *DSM-II* to public attention in 1973. It was still very much on Spitzer's mind while he was developing the *DSM-III*. On one side, many analysts remained firmly committed to the position that people with homosexual preferences had a psychiatric disturbance that resulted from deeply rooted, unresolved conflicts originating in early childhood.[73] They also believed that psychoanalytic treatments could uncover and resolve these problems. They worried that Spitzer's 1973 revision, which limited the diagnosis to those homosexuals who were distressed by their condition, could preclude insurance coverage for what they regarded as demonstrably effective treatments.

Spitzer responded that distressed homosexuals could still be diagnosed with sexual orientation disorder, which applied only to people who were disturbed by their sexual proclivities.[74] He developed the neologism "homodysphilia" to refer to this class. Spitzer's new category still assumed that homosexuality could be a mental disorder, despite the distress caveat. Prominent analyst Judd Marmor succinctly pointed out to Spitzer the absurdity of his proposal: "Either homosexuality in and of itself is a mental disorder or it is not."[75] Richard Green, a leading gay activist and member of the advisory committee on Psychosexual Disorders, expressed outrage at Spitzer's proposal and resigned from the committee.[76]

The ultimate result was a category of "ego-dystonic homosexuality," which applied only to individuals who were distressed at their lack of heterosexual arousal and presence of sustained patterns of homosexual arousal.[77] This diagnosis subtlety changed the focus from distress over homosexual inclinations toward impairment in heterosexual functioning. Some prior critics, including Marmor, were persuaded by this shift; others, such as Green, remained adamantly opposed. Yet all parties were exhausted by this nearly decade-long conflict, and the issue dissipated for the time being.[78]

Posttraumatic Stress Disorder

PTSD was the *DSM-III*'s most anomalous diagnosis. The dynamics surrounding the creation of this condition did not resemble those of the construction of any of the manual's other diagnoses. The experiences of military psychiatrists in World War II had led *Medical 203* and the *DSM-I* to include a diagnosis of gross stress reaction in response to intense environmental traumas. It emphasized how this condition was both temporary and reversible. The *DSM-II* dropped the diagnosis and replaced it with one called "transient situational disturbance." A subcategory of this condition used an example: "Fear associated with military combat and manifested by trembling, running and hiding," which could discourage traumatized combatants from seeking treatment.[79] In addition, no stress-related diagnosis recognized that responses to traumas could be chronic as well as transient and delayed rather than immediate.

Notably, PTSD was the sole *DSM-III* condition that emerged as a direct result of demands from outside interest groups. Shortly after the publication of the *DSM-II* in 1968 an energized group, Vietnam Veterans Against the War (VVAW), and allied psychiatrists demanded a diagnosis that incorporated long-lasting or long-delayed traumatic symptoms that resulted from combat-related stressors. By the mid-1970s so much time had passed since most Vietnam veterans had served that the existing definition excluded their prolonged suffering. Advocates created highly publicized media portrayals of large numbers of psychically traumatized veterans who were unable to obtain psychiatric treatment.[80]

Spitzer initially resisted the veterans' demands. One reason was that the diagnosis of "post-Vietnam syndrome" they pushed contained symptoms such as atonement, scapegoating, and "hatred of Orientals" that were far afield from traditional psychiatric criteria.[81] Veterans also emphasized the importance of their lived experiences, which the *DSM-III* tried to avoid. Moreover, a condition that by definition arose because of extreme social stress contravened the manual's elimination of etiology as a basis of classification. Finally, some task force members emphasized that PTSD was almost always accompanied by symptoms of anxiety, depression, or personality disorder and so very rarely arose in isolation. Consequently, it did not seem to be a free-standing condition.

Spitzer thus faced a dilemma. The controversy over homosexuality had led him to be highly attuned to media-savvy pressure groups. He also realized both that veterans were a sacred group in American society and that sympathy for antiwar veterans had deeply penetrated the APA.[82] When psychiatrists aligned with the

VVAW approached him about developing a stress-related diagnosis, he appointed the two most prominent collaborators with that group, Robert J. Lifton and Chaim Shatan, along with a Vietnam veteran, to the six-member Reactive Disorders advisory committee. They convinced the other members, one of whom was Spitzer himself, to add the PTSD diagnosis.

Despite the absence of any of the data-driven trappings that attended most of the other major *DSM-III* diagnoses, the reactive disorders committee developed a set of empirical criteria for the new diagnosis. It had three fundamental components. The first was a stressor criterion that required the "existence of a recognizable stressor that would evoke significant symptoms of distress in almost everyone." This fully embraced veterans' concern over displacing any responsibility for symptoms from the environment to the individual. The text accompanying the diagnostic criteria specified that qualifying traumas are "outside the range of usual human experience," providing a unifying framework for all kinds of severe traumas including not only military combat but also rape, assault, and natural or man-made disasters.[83]

The symptom criteria for the diagnosis required, first, that the trauma be reexperienced through intrusive recollections, recurrent dreams, or feelings of reoccurrence. Second, sufferers had to show numbed responsiveness through diminished interest in activities, detachment from others, or nonreactive moods. Finally, they had to display at least two symptoms from among hyperalertness, sleep disturbance, survivor guilt, trouble concentrating, avoidance of reminders of the traumatic event, and intensification of symptoms following such reminders.

Last, in addition to acute PTSD involving symptoms that arose within six months of the trauma and disappeared within this time frame, the criteria allowed for chronic PTSD that either persisted for longer than six months or arose more than six months after the trauma. This aspect of PTSD responded to veterans' need for a diagnosis that encompassed persistent or delayed symptoms.

In one of the ironies of psychiatric history, the PTSD criteria that specify the presence of "recurrent and intrusive recollections of the event; recurrent dreams of the event; and sudden acting or feeling as if the traumatic event were reoccurring" closely resembled Freud's view of trauma that emerged from World War I.[84] While the manual almost completely purged Freudian notions, they remained unacknowledged in the PTSD diagnosis. Despite the fact that this condition was a thoroughgoing anomaly in the manual, it would go on to become, with MDD, the most significant diagnosis in the entire *DSM*. Once the manual recognized long-delayed and chronic, as well as acute, forms of PTSD, many patients clamored for a label that could secure them substantial benefits.[85]

Substance Use Disorders

The empiricist nature of the *DSM-III* is especially apparent in how its definitions of substance use disorders diverged from those of the prior manuals. The *DSM-II* quickly disposed of this category, which it placed under the class of "other non-psychotic mental disorders," in a page and a half. It described alcoholism as "for patients whose alcohol intake is great enough to damage their physical health, or their personal or social functioning, or when it has become a prerequisite to normal functioning." This brief definition is followed by equally short descriptions of three subtypes of episodic excessive drinking, habitual excessive drinking, and alcohol addiction. The manual cursorily defined drug dependence as "habitual use or a clear sense of need for the drug" and provided a list of relevant drugs.[86]

In contrast, the *DSM-III* contained a general class of substance abuse and substance dependence that it divided into alcohol abuse and dependence and drug abuse and dependence. It devoted 16 pages to criteria for the category, leaving no doubt about how to define abuse and addiction. The manual further divided each condition into many different manifestations. For example, diagnostic criteria for alcohol abuse note nine indications of pathological use, five types of impairment in functioning due to use, and a duration criterion of at least one month. Because the earlier two manuals made no etiological assumptions about substance use and abuse, the *DSM-III* had no need to change this aspect of the category.[87]

Childhood Disorders

The class of what the *DSM-II* called "behavior disorders of childhood and adolescence" and the *DSM-III* renamed "infancy, childhood, or adolescence disorders" also illustrates the differences between the two manuals. The *DSM-II* situated these conditions very generally as "more stable, internalized, and resistant to treatment than *Transient situational disturbances* but less so than *Psychoses, Neuroses,* and *Personality Disorders.* This intermediate stability is attributed to the greater fluidity of all behavior at this age."[88] It then spent a page providing brief definitions of seven types of mental disorder among youth. In contrast, the *DSM-III* devoted a hefty 64 pages to its comparable category, almost double the size of all the definitional material in the entire *DSM-II.* It then provided specific definitions for 31 discrete conditions. Their placement at the very front of the manual heralded the vastly greater attention that future *DSM*s would pay to this class.[89]

Autism provides an example of the *DSM-III*'s changing awareness of childhood conditions. The first two manuals mentioned "autism" only as a symptom of childhood schizophrenia. The *DSM-III* moved autism from the schizophrenic category to the chapter on disorders usually first evident in infancy, childhood, or adolescence. It provided two specific criteria sets for what it called "pervasive developmental disorders" that could emerge before 30 months of age or between 30 months and 12 years of age. These diagnoses clearly distinguished infantile and childhood-onset autism from any psychoses, realized that they often arose at an early age, and recognized the pervasive and severe impairments they usually entailed. The *DSM-III* provided the foundation for the vast expansion of awareness and treatment for a serious condition beginning in childhood that previous manuals had ignored.[90]

The Impact of the *DSM-III*

In contrast to the release of its unassuming predecessors, the publication of the *DSM-III* was celebrated as a major event in American psychiatry, immediately propelling the manual to recognition in the broader culture.[91] The various disputes involved in its creation were forgotten, as was the shaky evidence base for its diagnostic criteria. The public soon came to accept the manual's basic principle that mental disorders were specific entities that closely resembled physical diseases. Sociologist Jason Schnittker's Google nGram analyses show that references to the *DSM* were virtually nonexistent before 1980, when they sharply escalated and continued to expand until leveling off around 2000. In addition, they indicate a significant change in how the public viewed mental illnesses: after 1980 it increasingly considered mental disorders something independent of individuals (e.g., "have depression") as opposed to something that is an individual attribute (e.g., "am depressed"). Schnittker's research also provides evidence for an explosion of references to the *DSM-III*'s specific disorders in fictional narratives.[92]

Within a short time, the *DSM-III*'s triumph within psychiatry was complete. Despite the intense conflicts that accompanied its adoption, the *DSM-III* quickly gained widespread acceptance within the profession. Almost all nonanalytic psychiatrists agreed with Robert Kendell's pronouncement that "it is far superior to any previous psychiatric classification."[93] Psychologists and other mental health professionals, too, accepted the manual's authority over diagnoses of mental illness. In 1984, after noting the hegemony it had attained in psychiatry, Gerald Klerman flatly stated, "This debate is already an anachronism. The victory of DSM-III has been acknowledged by our colleagues and adversaries in psychology, in the other mental health professions and in other countries."[94]

The *DSM-III* immediately became the standard for training in psychiatric education. All major psychiatric journals expected that submissions use the *DSM* categories. By 1990 more than 2,300 scientific articles referenced the manual.[95] Likewise, the NIMH grounded its research program in the *DSM-III* diagnoses. Clinicians, too, quickly recognized the manual's value for them. A survey published in 1981 indicated that securing payments from third parties was their primary reason for using the *DSM*.[96]

Initially, psychoanalysts were appalled by the new manual. In their eyes, the task force was composed of a small group of people who had "such an arrogant view of their mission and are not willing to incorporate some of the things which we have learned over the past 70 years."[97] Prominent analytic psychiatrist George Vaillant mocked the manual's reliance on symptoms: "It is as if cough were not distinguished from pneumonia."[98] Analysts recognized, however, that their era of dominance within the profession was over, and their earlier fervent critiques withered away. By now, psychoanalysis has such a tiny presence within psychiatry that it is no longer a subject worth discussing.[99]

The *DSM-III* was accurately seen as a "revolution" in psychiatric diagnosis. It also seemed to herald the triumph of the scientific viewpoint in psychiatry. A popular 1985 book entitled *The New Psychiatry* summarized, "There has been a revolution in psychiatry. In the 1980s, mainstream American psychiatry has switched from being primarily psychoanalytic to being primarily scientific." Its author, psychiatrist Jerrold Maxmen, was unambiguous about the value of this transformation. The old psychiatry stemmed from theory—"something is true because Freud said so." In contrast, the new psychiatry derived from facts that are the result of scientific procedures. Treatments, too, markedly differed. "Ideology primarily directs the psychoanalytic psychiatrist's choice of treatment, whereas pragmatism primarily directs that of the scientific psychiatrist."[100] While there is much room to debate the scientific credentials of the *DSM-III*, there is no question that its developers believed it was a major scientific achievement.[101]

It took a long time—about two decades—before psychiatric researchers recognized the serious problems in the *DSM-III*. Their findings indicated that the basic characteristics of mental disorders were virtually the opposite of those portrayed in the manual: overlapping rather than discrete, dimensional more than categorical, generalized and not specific, and reflective of a small number of basic processes as opposed to hundreds of distinct conditions. Before these aspects would penetrate the *DSM* process, however, it would go through two further revisions.

FIVE

The *DSM-III-R* and *DSM-IV*

T he *DSM-III* almost immediately gained hegemony over practice, research, and teaching in psychiatry and other relevant disciplines. The manual dictated the acceptable scope of thinking about mental disorder as clinicians and research-ers alike, psychiatrists of all theoretical perspectives, and practitioners from every major mental health discipline accepted the new classification. The *DSM-III* was also quickly institutionalized among medical professionals, government bureaucrats, hospital administrators, mental health educators, advocacy groups, pharmaceutical companies, the insurance industry, and the judicial system. Pa-tients, too, acquired a new language to interpret their distressing experiences and explain their emotional lives. They also needed a *DSM* diagnosis to be reimbursed for their treatment. Perhaps most important, after undergoing almost two decades of sustained criticism, the *DSM-III* restored the prestige of the psychiatric profes-sion within both medicine and the general culture.[1]

The omnipresence of the *DSM* diagnoses did not entail that its adopters truly believed the manual captured the essential nature of mental disorder. Clinicians, in particular, were often skeptical about the validity of many particular diagno-ses: "Doctors don't care much about diagnosis. They use diagnosis mostly for codes [to obtain reimbursement]. They don't really care what the rules are," prominent psychiatrist Michael First observed.[2] Government programs such as Medicaid and Social Security Disability Insurance also mandated that mental health care organizations use *DSM* diagnoses to receive funding. As one community mental health worker put it, the manual became "our little book for witchcraft so we can bill."[3] Despite criticisms that the new manual was overly empiricist, reductionist, and ignorant of psychiatric history, the success in implementing the *DSM-III* far exceeded the expectations of Spitzer and his allies.[4]

The framework established in the *DSM-III* remained intact in the *DSM-III-R* (1987) and *DSM-IV* (1994). Both manuals embodied a standard medical model that they justified according to empirical findings. Like the *DSM-III*, the next two editions used explicitly defined syndromes based on descriptions of observable symptoms, fairly precise inclusionary and exclusionary criteria, discrete diagnostic categories, and theory neutrality. Each maintained the bedrock principle that making progress in explaining and treating mental disorders depended on identifying and accurately describing specific diagnostic entities.

Given that the subsequent two versions would maintain the framework established in the *DSM-III*, what generated these revisions? The American Psychiatric Association invoked the same rationale it used for *DSM-II* and *III*: the need for American diagnoses to be congruent with the World Health Organization's *International Classification of Diseases*. This justification warrants skepticism, given the extensive differences that remained between the *DSM* and the *ICD* after each revision. The APA also emphasized the need to incorporate new research findings into the revised manuals, although neither makes extensive use of such evidence.

Another motivation is likely the huge profits the APA reaped from each revised manual.[5] The immediate acceptance of the *DSM-III* produced an unexpected financial windfall for the APA. Spitzer noted how the *DSM-III* "made an unbelievable amount of money for the APA. That was a huge surprise."[6] The *DSM-III-R* sold more than a million copies, generating nearly $20 million of revenue. The *DSM-IV* averaged $6.5 million in yearly sales.[7] The APA also published a *DSM-IV-TR* (text revision) in 2000 that was basically a reprinting of the *DSM-IV* with a few minor changes and corrections in particular criteria sets.[8] The *DSM* became a financial pillar of the APA, but it required revisions to profitably generate new markets.

Spitzer maintained control of the *DSM-III* revision that began in 1983. "The organizational principle that brought cohesion to *DSM-III* and *DSM-III-R* was the omnipresent leadership of Bob Spitzer, who chaired every work group, nursed every detail, and wrote every word," Spitzer's successor in the *DSM-IV* process, Allen Frances, observed.[9] Spitzer headed the eight-person work group charged with revising the manual as well as all 26 advisory committees to the work group, and he appointed more than 200 consultants. The *DSM-III-R* added a considerable number of new diagnoses and modified many existing diagnostic criteria.[10] Almost all of these changes made it easier to obtain a psychiatric diagnosis; no central types were eliminated. The result was that *DSM-III-R* contained 292 diagnoses, compared to the 265 diagnoses found in the *DSM-III*. The general classes established in the *DSM-III* remained intact, with the exception that the revision added a group of sleep disorders, with 12 different diagnoses.[11]

The intensely public, and often embarrassing, debates (discussed below) that surrounded the *DSM-III-R* led the APA to try to avoid similar controversies in the next revision. Because Spitzer had become such a polarizing figure, the APA replaced him with Frances, who was more clinically and analytically oriented than his predecessor was, to lead the *DSM-IV* process.[12] Frances appointed a more diverse group of members and consultants in terms of theoretical allegiances, types of practice settings, gender, and ethnicity.[13] He also decentralized the development of this revision to a far greater extent. Frances strove to be far more conservative than Spitzer, seeing his role "as one of refiner/conservator rather than innovator. . . . A change would be made only when it was compellingly necessary and there was an overwhelming consensus on the science." The *DSM-IV* revision promoted a more restrained approach, where diagnostic changes would be "few and far between." It contained only five more total diagnoses than the *DSM-III-R* and only eight new diagnoses, far fewer than the number of diagnoses added in *DSM-III-R*. The result was that "*DSM-IV* was faithful to *DSM-III-R*."[14]

The *DSM-IV*'s guiding principle was to require strict standards of evidence for introducing new diagnoses. Its various work groups were charged with compiling the extensive literature reviews and meta-analyses of the manual's most important diagnostic categories.[15] Although they produced four hefty source books that contained a massive amount of data, this information justified the existing conditions and assumed that they had already passed the evidentiary bar; there was little relationship between the materials found in the source books and the modest changes that entered the *DSM-IV*.[16] Despite its reviews, data reanalyses, and field trials, the manual made no basic modifications in preexisting criteria. In effect, the *DSM-IV* simply ratified the existing structure of psychiatric diagnoses. Spitzer presciently forecast a year before the publication of the *DSM-IV*, "My own prediction is that when final decisions are made about DSM-IV, they will still be based primarily on expert consensus, rather than on data, as was the case with the DSM-III and DSM-III-R."[17]

The most significant adjustment in the *DSM-IV* was to abandon the separation of organic mental disorders from other general categories, which had been the core principle of the *DSM-I* and persisted as a stand-alone class in the following three manuals. "The terms 'organic' and 'nonorganic' diseases have been eliminated on the grounds that they imply a mind/body dualism and that some mental disorders are not true medical illnesses," a *JAMA* commentary documenting this shift observed.[18] The *DSM-IV* itself noted that "mental" and "medical" are "merely terms of convenience and should not be taken to imply that there is any fundamental distinction between mental disorders and general medical conditions, that mental disorders are unrelated to physical or biological factors or processes, or that general medical conditions are unrelated to behavioral or psychosocial factors or processes."[19] The abolition of a separate organic category, which implied that *all* of the *DSM* mental disorders were at least partly organic, was the final nail in the coffin of the psychoanalytic approach to diagnosis.

Although the APA removed Spitzer from the *DSM-IV* process, the manual remained true to his original vision of symptom-based, empirical, discrete, and theory-neutral diagnoses. It considered using continuous measurements that have no distinct cutoffs between disorder and non-disorder for some conditions but rejected the idea because, "unfortunately, there are not any dimensional systems either for Axis I or for Axis II that have achieved acceptance, and the dimensional model appears too unsupported by compelling evidence and is too radically innovative to be considered seriously for inclusion in *DSM-IV*."[20] Although some new diagnoses appeared in the *DSM-III-R* and *DSM-IV*, almost all of the diagnoses that emerged in the *DSM-III* remained. The major difference was that, while intraprofessional concerns almost completely dominated the making of the *DSM-III*, outside interest groups played more important roles in shaping the changes in the *DSM-III-R* and *IV*.

The *DSM* Escapes Professional Confines

The central changes in the *DSM-III-R* and *IV* processes were not diagnostic. Instead, the major transformation in the *DSMs* after 1980 was their emergence as important cultural documents. With the exception of the antagonistic debates over the homosexuality issue in the early 1970s, the *DSM-I* and *II* had been socially invisible. The development of the *DSM-III* attracted considerable public interest, but the manual's particular diagnoses were not well known outside of professional practice. During the 1980s, however, the *DSM* metamorphosed from a manual that served the purposes of a professional guild to one that became embedded in public consciousness.

Diagnoses, which once centered on the labels that clinicians applied to patients, became matters of cultural scrutiny, shapers of self-awareness, and generators of massive profits. After 1980, references to the *DSM* exploded, increasing exponentially every year until the turn of the century.[21] The manual also came to frame portrayals of mental illness in television programs, popular magazine articles, autobiographies, and science reporting to the general public. Diagnostic changes proposed for the *DSM-III-R* and *IV* attracted intense media attention. Perhaps most important, as discussed below, *DSM* diagnoses shaped the promotion of all psychotropic drugs that entered the market after 1980.

A second notable change was the greater input from external interest groups in the two *DSM* revisions. In contrast to the almost exclusively intra-professional discussions that produced the *DSM-III*, many outside advocates were involved in the *DSM-III-R* and *DSM-IV* revisions. Not just psychiatrists but other mental health professionals, mental health advocacy organizations, and feminist groups all engaged in intense lobbying efforts. While the drug industry capitalized on the *DSM* system, it did not play a direct role in constructing or modifying any diagnosis.

Another new development that accompanied the revisions to the *DSM-III* was the entry into public consciousness of the epidemiology of mental illness. Epidemiology is the study of amount of disease in the general population, not in samples of treated cases. Earlier community studies, such as the Midtown Manhattan Study (1962), were unable to assess rates of specific mental illnesses because no empirical measures existed at the time. Instead, population research before the *DSM-III* examined more general psychic difficulties among putatively normal people. The *DSM-III*'s descriptive criteria sets allowed epidemiologists to estimate rates of particular mental disorders in untreated groups for the first time.[22]

Epidemiological studies unexpectedly found that extraordinarily large numbers of people seemingly had at least one *DSM* disorder. About half the population met criteria for some disorder in their lifetime and more than a quarter in a given year. Nearly a third reported some lifetime anxiety disorder, over a fifth a mood disorder, and a quarter some impulse control disorder.[23] Almost 15 percent met criteria for at least one personality disorder.[24] Moreover, broadening diagnostic criteria from the *DSM-III* to the *DSM-III-R* and *IV* led to steady increases in the rates of putative mental disorders. The enormous estimates that these studies produced became powerful marketing tools for psychiatry and other mental health professions, mental health advocacy groups, the NIMH, and, especially, the drug industry. The application of the DSM diagnoses to the whole of society seemed to indicate that virtually everybody, not just psychiatric patients, suffered from a mental disorder at some point in life.

The intense debates over psychiatry's deficiencies in the 1960s and 1970s revolved around general issues such as the very existence of mental illness, the justifiability of involuntary mental hospitalizations, and the legitimacy of the psychiatric profession itself. In contrast, discussions after 1980 centered on the merits or deficiencies of particular DSM diagnostic categories. While earlier disputes had focused on the psychoses and particularly schizophrenia, after 1980 the most visible diagnoses became those that either had implications for outside interests or high marketing potential for drug companies.[25]

Feminist Opposition to Diagnoses: PMS and SDPD

One notable aspect of the DSM-III-R and IV processes was the heightened involvement of feminist groups. Although men had always accounted for the bulk of hospitalized populations, women traditionally made up about double the number of psychiatric outpatients.[26] Almost all their entries were self-initiated. As rates of involuntary commitments to institutions rapidly diminished and those of voluntary treatment rose exponentially over the latter half of the twentieth century, DSM labels were increasingly applied to the problems that women brought to therapy. This feminization of psychiatric diagnoses led the women's movement to scrutinize the sorts of conditions they applied to more intensely.

Feminist groups led both the opposition to and the advocacy for varying diagnoses. Their questioning of the validity of menstruation-related conditions and personality disorders related to feminine traits led the revised manuals to abandon or weaken these diagnoses or diagnostic proposals. Conversely, these groups were major proponents for resurrecting the diagnosis of multiple personality disorder (MPD) and expanding the range of posttraumatic stress disorder. Feminists were the most visible among the variety of interest groups that became involved in the DSM revision processes.

Although feminist psychiatrists had voiced concerns over what they viewed as stigmatizing diagnoses in the development of the DSM-III, their objections were less visible than the disputes over the neuroses, PTSD, and the personality disorders. Feminist objections, however, dominated the DSM-III-R revision. The most divisive controversy in this process was the very public critiques of the task force's efforts to develop more systematic criteria for the concepts of premenstrual syndrome and of masochistic personality, which the manual called late luteal phase dysphoric disorder (LLPDD) and self-defeating personality disorder (SDPD), respectively.

One of the 26 advisory committees that drafted the *DSM-III-R* considered the possible introduction of a new diagnosis concerning the mood swings that many women experienced around the time of menstruation.[27] This diagnosis, which involved symptoms such as affective lability, irritability, anger, and depression, generated considerable controversy. Psychiatrists affiliated with the APA's Committee on Women enlisted opponents from a number of other professional organizations and lay feminist groups to join their fight against what was initially called premenstrual dysphoric disorder (PDD). In contrast to the evidence-based arguments of the researchers on the advisory committee, feminist opposition was focused on the social consequences of a menstruation-related diagnosis. They asserted that "the [PDD] diagnosis has ominous implications for perpetuating damaging stereotypes of women and for fostering an unnecessarily pathological view of women's experience."[28] Feminists replaced psychoanalysts as the most vocal opponents of the *DSM-III-R* process, which spilled into the popular press and generated protests at APA meetings.[29]

The ad hoc committee considering the PDD diagnosis, which contained eight women and six men, including Spitzer, was well aware of feminist opposition to calling PMS a psychiatric disorder. Their discussions, however, focused on empirical issues such as what particular symptoms to include in the criteria set, how to differentiate PDD from other diagnoses such as dysthymia and depression, and whether the diagnosis should include physical, psychological, or both types of symptoms.[30] At first, Spitzer insisted that the manual must include what was briefly called "periluteal phase dysphoric disorder": "When psychiatry is under attack, you must not equivocate. There is no doubt that what we call periluteal phase dysphoric disorder (the newly adopted name for what had been called premenstrual dysphoric disorder) exists as a clinically significant syndrome in some women, even though the causes and optimal therapy are unknown. We must call this a mental disorder!"[31] As with the *DSM-III* controversies, however, Spitzer came to realize that compromise was a better tactic than unequivocal opposition.

Spitzer asked the ad hoc committee on PMS to gather additional data to respond to the feminist critique. In response, it modified the diagnosis; created a higher threshold for meeting the criteria, including the presence of "marked" impairment; and ruled out a PMS diagnosis when the patient met criteria for any other diagnosis. The committee also broadly circulated its proposed new criteria to authors of relevant publications on PMS. The committee then recommended including PDD in the main text of the manual.[32]

The issue of whether the *DSM-III-R* should include a diagnosis of self-defeating personality disorder was equally controversial. Psychologist Paula Caplan became the most visible opponent of diagnoses perceived to discriminate against women. Caplan believed that the SDPD diagnosis reflected social expectations that women put other people's needs ahead of their own, rather than constituting a mental disorder. She also feared that the criteria for SDPD, such as choosing to remain in relationships involving maltreatment, rejecting attempts at help, and refusing opportunities for pleasure, typified the behaviors that female victims of violence displayed.[33] In effect, women in these situations would be doubly victimized, first by their violent male domestic partners and then by psychiatrists. Caplan engaged in widespread media campaigns, wrote articles for the popular press, and appeared on a number of popular television programs deriding the proposal for a diagnosis related to women's masochism.

In response to the feminist opposition to SDPD (then called "masochistic personality disorder"), the advisory committee on Personality Disorders met with members of feminist psychiatric and psychological groups. A representative of the Feminist Therapy Institute told the committee that her organization "had voted to invest all its financial and other resources in filing a lawsuit" if it approved this diagnosis and the associated one of PDD.[34] This meeting generated considerable adverse publicity in the popular press regarding the *DSM-III-R* process. In response, the advisory committee changed the label of masochistic personality to self-defeating personality disorder, which did little to mollify the feminists, who considered both labels to be victim blaming. They also noted how the *DSM* contained no parallel label for traditionally masculine behavior.[35] Groups aligned with the opposition expanded beyond the two APAs to include the National Association of Social Workers, National Association of Women and the Law, the American Orthopsychiatric Association, and a number of other organizations. Echoing the protests surrounding the homosexuality diagnosis in the *DSM-II*, these groups organized highly visible and widely publicized demonstrations at the 1986 APA meeting.

The central question regarding these diagnoses was whether including them in the *DSM* would discriminate against women, as feminist groups argued, or help them gain the treatment they desired, as Spitzer and the majority of the advisory committee members believed. After all, the committee reasoned, no one was *im-*

posing any diagnosis on anyone; women who received the supposedly objectionable diagnoses were ones who sought psychiatric help for their problems. Psychiatrists weren't telling them, as Caplan insisted, that they were crazy.[36]

In the end, the APA took a third position. It was less concerned with either rejecting a stigmatizing diagnosis or accepting a helpful one than with avoiding reputational damage to the psychiatric profession. In a surprising rebuke of Spitzer and its own advisory committees, the APA's Board of Trustees voted to relegate LLPDD to an appendix called "proposed diagnostic categories needing further study." In addition to LLPDD, the appendix contained criteria for self-defeating personality disorder. Sensitive to feminist objections, it balanced SDPD with a new category of sadistic personality disorder, which was "far more common in males than in females."[37] The stated goal was to allow more research data to accumulate that would either support or reject the inclusion of each of these three conditions in the main text of a future manual.[38] In fact, placing these diagnoses in an appendix was a deft political compromise.[39]

An unintended consequence of the PMD and SDPD controversies was the ouster of Robert Spitzer from the leadership of the *DSM* process. The APA was distressed over the negative publicity the debates generated and blamed Spitzer for allowing too many nonpsychiatrists into the process. They wanted to ensure that the next revision of the *DSM* would remain out of the public eye to avoid the embarrassing protests that marked the development of the *DSM-III-R*.[40]

The APA's attempt to avoid further controversy failed. Indeed, antagonism to the LLPDD diagnosis intensified in the *DSM-IV* deliberations.[41] Caplan was again its most visible opponent, appearing on an episode of *The Phil Donahue Show* titled "Psychiatrists Want to Classify Women with PMS as Crazy." In a joint appearance with Spitzer on *The Today Show*, Caplan proclaimed that "this really should properly be called PMSgate because of the history of lies and distortions that have characterized the whole process. The APA wouldn't dream of trying to classify half a million Black people or gays as mentally ill with no research to support it."[42] Numerous media reports in publications including the *New York Times*, the *Washington Post*, the *Wall Street Journal*, and *Newsweek* covered this debate. The result was that SDPD (and SPD) were removed from the *DSM-IV* while LLPDD, renamed "premenstrual dysphoric disorder," remained in an appendix for diagnoses awaiting further study.[43] The potential professional discomfiture from disorders that politically organized groups considered to be stigmatizing ultimately trumped the presumed scientific basis for these diagnoses.

Feminist Support for Expanding Diagnoses: MPD and PTSD

Ironically, feminists were also central to a second, diametrically opposite, dynamic that strove to apply trauma-related diagnoses to a far broader range of women. In the early 1980s, as memories of prior wars faded, views of trauma as the psychological consequence of violence that men inflicted on women became a significant theme for some feminists.[44] A backlash had developed against sexual liberation, a prominent issue in the feminist movement during the 1960s and 1970s, emphasizing the rights of women to reproductive choice and the same erotic freedoms as men. In the 1980s, many leading feminists, including Catherine McKinnon, Andrea Dworkin, and Gloria Steinem, vocally expressed their concerns that the advocacy for sexual choice in recent years had resulted in even more male oppression over women. They focused on the dire impacts for women of family violence, date rape, workplace sexual harassment, and childhood sexual abuse. "Perhaps incestuous rape," Dworkin proclaimed, "is becoming a central paradigm for intercourse in our time."[45]

During the 1980s the recovered memory movement (RMM) gained prominence, emphasizing the prevalence of repressed, and then recovered, memories of childhood sexual abuse.[46] The basic tenet of the RMM, echoing Freud's early, quickly repudiated views and Pierre Janet's turn-of-the-century theories about dissociation, was that repressed memories of early traumas in childhood remained present in an unconscious state. "The ordinary response to atrocities," according to psychiatrist Judith Herman, a leader of the RMM, "is to banish them from consciousness."[47] The repression of traumatic experiences bore little resemblance to the intense memories of veterans, who were all too able to recall their traumas. It also contradicted a vast body of evidence indicating traumas were accompanied by highly intrusive and often terrifying images that gradually fade with the passage of time.[48] Nevertheless, a large cadre of therapists came to be committed to techniques that purported to bring long-repressed memories into consciousness. This group required diagnostic criteria that recognized that traumatic experiences could be repressed as well as intrusive.

Initially, mental health professionals allied with the RMM resurrected the long-neglected condition of multiple personality disorder to diagnose their patients. This condition entered public consciousness through Flora Rheta Schreiber's bestselling *Sybil: The True Story of a Woman Possessed by 16 Separate Personalities*(later made into a highly successful movie), which sold 6 million copies in the United States alone. Before its 1973 publication, MPD diagnoses were extremely rare. Although the character that this nonfiction book was based on was later unmasked as a fraud, *Sybil* caused the diagnosis to skyrocket.[49]

Neither the *DSM-I* nor *II* contained a diagnosis related to MPD, but the *DSM-III* placed the new condition in the class of dissociative, not personality, disorders. The essential feature of this group was "alteration in the normally integrative functions of consciousness, identity, or motor behavior." The manual defined multiple personality as "the existence within the individual of two or more distinct personalities, each of which is dominant at a particular time." It added that "child abuse and other forms of severe emotional trauma in childhood may be predisposing factors." However, the *DSM-III* also indicated that "the disorder is apparently extremely rare."[50]

The *DSM-III-R* made some small changes in the diagnostic criteria for MPD, noting "that in nearly all cases, the disorder has been preceded by abuse (often sexual) or another form of severe emotional trauma in childhood." Moreover, the manual noted that "the disorder has been diagnosed from three to nine times more frequently in females than in males." It further indicated, "Recent reports suggest that this disorder is not nearly so rare as it has commonly been thought to be."[51] Indeed, between 1985 and 1995, what the *DSM-III* had called an "extremely rare" condition generated almost 40,000 new diagnoses. MPD also gained social prominence as numerous daytime television programs featured guests with this highly telegenic condition and as celebrities such as Roseanne Barr claimed to have it.[52]

However, the diagnosis was quickly discredited as sufferers became increasingly likely to recollect wildly implausible childhood traumas including satanic murders of babies, outlandish rapes, and multiple incestuous abusers. A number of parents who were targets of their children's accusations, and sometimes patients themselves, successfully sued the clinicians they claimed had induced their condition. In one highly publicized case, psychiatrist Bennett Braun, a member of the *DSM-III-R* Dissociative Disorders advisory committee, settled one of his patient's lawsuits against him for nearly $11 million. While in therapy, this patient claimed she was a priestess in a satanic cult that had, among other things, cannibalized 2,000 children a year and watched as she was raped by gorillas, panthers, and tigers at a zoo.[53]

The media, which initially widely publicized this diagnosis, took heed of legal developments and became increasingly skeptical of MPD. Likewise, insurance companies stopped reimbursing long-term treatments for the condition. MPD had become an embarrassment to the psychiatric profession. The *DSM-IV* eliminated the diagnosis, replacing it with a reformulated condition of dissociative identity disorder (DID). According to the head of the relevant *DSM-IV* work group, this change occurred because "we wanted this condition to be regarded like any other mental disorder, and not like some weird, far-out, cultlike thing."[54] The new DID diagnosis, however, remained tainted by its earlier incarnation as MPD. After the late 1980s, PTSD became the preferred diagnosis for symptoms that resulted from all sorts of traumas.

Before the First World War, women's traumas had been at the heart of psychiatric attention, playing a central role in the development of psychoanalytic theory.[55] The war, however, turned concern away from childhood traumas among women to the terrors of male combatants. Wartime traumas were also the basis for the central DSM-I diagnosis of gross stress reaction, designed for "previously more or less 'normal' persons who have experienced intolerable stress."[56] The DSM-II, however, had no comparable stress-related diagnosis. As chapter 4 discussed, the vigorous advocacy efforts of veterans of the Vietnam War and allied psychiatrists led to the reappearance of a traumatic diagnosis, posttraumatic stress disorder, in the DSM-III.

The PTSD criteria specified the "existence of a recognizable stressor that would evoke significant symptoms or distress in almost everyone" as a necessary component for a diagnosis. They thus clearly located the origins of PTSD in a major environmental stressor rather than in any individual vulnerability. While the diagnosis was primarily developed to apply to the psychic wounds stemming from military combat, the accompanying textual description also mentioned other extreme stressors such as rape and natural and man-made disasters as possible progenitors of PTSD.[57]

Although feminist clinicians and advocates had little involvement in creating the DSM-III PTSD criteria, once the diagnosis appeared they saw its relevance to their concerns: traumatized women who were victims of male violence and abuse. PTSD appealed to feminist groups because, unlike the SDPD diagnosis, it clearly located the source of psychic problems in the environment, not within individuals. "The women's movement greeted PTSD enthusiastically because it created a diagnostic niche for victims of rape, domestic violence, child abuse, and sexual assault," Christina Sommers and Sally Satel observed.[58] As the focus of traumatic attention shifted, "the most common post-traumatic disorders are those not of men in war but of women in civilian life," Judith Herman proclaimed in her canonic book, *Trauma and Recovery*.[59]

PTSD was again singular in the DSM-III-R and DSM-IV. The overwhelming number of DSM-III diagnoses remained unchanged or underwent small alterations in the two revisions. In contrast, the 1987 and 1994 revisions substantially modified the PTSD diagnosis. Spearheaded by feminist advocates, both manuals significantly expanded who could meet the PTSD criteria.

The major revisions in the *DSM-III-R* rendered PTSD diagnoses more suitable for clinicians who were concerned with the repressed traumas of their female patients. Primarily in response to the recovered memory movement, the *DSM-III-R* criteria changed the category of numbed responsiveness to the external world to include the "inability to recall an important aspect of the trauma." Symptoms now encompassed repressed as well as recurrent and intrusive memories of some trauma. The *DSM-III-R* also significantly lowered the duration requirement from six months to one.[60]

The *DSM-IV* brought about an even greater expansion of the PTSD diagnosis. The central principle of this manual was supposed to be conservative; it would change existing diagnostic criteria only in the face of overwhelming supportive evidence. Once again, the PTSD diagnosis was an exception. PTSD is a rare *DSM* diagnosis that uses etiology rather than descriptive phenomenology to define the condition: someone who has the defining symptoms without having experienced a relevant stressor cannot have the disorder. Therefore, broadening the criteria of exposure also increases the number of people who qualify for the diagnosis.

The *DSM-III* and *DSM-III-R* criteria had limited traumatic events to those that would lead to significant distress in "almost everyone," leaving little room for individual biological and psychological susceptibilities. The *DSM-IV* greatly enlarged the scope of the stressor criterion and thus the range of people who suffered from a qualifying trauma by dropping the requirements that events must "cause distress in almost everyone" and be "outside the range of normal human experience." It replaced them with the stipulation that "the person experienced, witnessed, or was confronted with events that involved actual or threatened death or serious injury, or a threat to the physical integrity of self or others," and that "the person's response involved intense fear, helplessness, or horror."[61]

The revised criteria did not eliminate any cases that the earlier definitions had captured. Instead, the altered definition both expanded the class of people who were considered to be exposed to traumatic stressors and increased the ambiguity regarding what constituted a traumatic experience. Indeed, the "confronted with" criterion extended the notion of exposure so that persons who were not even present at a traumatic event could potentially meet diagnostic criteria for PTSD. The stressor criteria were now heterogeneous enough to incorporate not just individuals who were directly involved in traumas but also those who only witnessed or were bothered by them. For example, someone who learned of the sudden and unexpected death of a close relative or friend who has died from natural causes or people who watched a disaster unfolding thousands of miles away on television could meet the new stressor criterion.

The *DSM-IV* diagnosis changed the nature of traumatic exposure in a second way. The mandate that the person's response involve "fear, helplessness, or horror" shifted the definitional criteria from the nature of the stressor to the experience of the victim. Individual temperament, personality, and emotional reactivity now entered into what defined an event as "traumatic" in the first place, introducing a subjective element into evaluating the stressor itself. The *DSM-IV* thus radically changed definitions of trauma both to include a great heterogeneity of experiences and to partially locate the nature of trauma within the individual rather than the environment.

These changes in the definition of who was considered to be exposed to a "trauma" had major consequences. The *DSM-IV* criteria for traumatic exposure were so expansive that they encompassed virtually everyone. Community surveys showed that about 90 percent of the population reported traumatic events that met the stressor criterion of the definition. This was about a 60 percent increase from those included by the previous definition. The most common traumatic event, reported by 60 percent of the population, was the sudden, unexpected death of a close relative or friend.[62] While learning that a loved one has died is surely a distressing experience, it is typically of a different order of magnitude than participating in combat or being sexually assaulted.

The boundaries of PTSD came to encompass ever-widening groups of claimants. The diagnosis was applied to a broad range of events that were far removed from those encompassed by the original definition. These included sexual and workplace harassment, exposure to hazardous materials, automobile accidents, crime victimizations, spousal infidelity, news about the death of intimates, or a terrorist attack broadcast on television. A new legal specialty of trauma law developed that provided assessment and diagnoses for people seeking compensation for the psychological damage from such injuries. Advocates, for example, argued that up to a third of sexual harassment cases led to PTSD symptoms, which should be "reflected in judicial decisions regarding the appropriate monetary compensations for victims of sexual harassment."[63]

While much of the discussion surrounding PTSD after the *DSM-III* involved escalating rates of the condition among male combat victims, the *DSM-III-R* and *IV* revisions had a considerably greater impact among women, who were more likely to experience all major categories of traumas: assaultive violence, injuries, exposure to others' traumas, and sudden, unexpected death.[64] Overall, 86 percent of women compared to 69 percent of men were exposed to these shocks. Given the experience of some traumatic event, women were also consistently more likely than men to develop PTSD.[65] A meta-analysis of gender differences in risk for PTSD among trauma-exposed adults "showed a consistent pattern for women to be at higher risk than men."[66] Another comprehensive literature review of 160 studies involving more than 60,000 victims of various disasters found that men were significantly more distressed in only 3, compared to 31 that showed more distress among women.[67] For example, in the aftermath of Hurricane Katrina in 2005, almost three times as many women qualified for PTSD diagnoses as men.[68] The continuing expansion of the PTSD diagnostic criteria thus elevated both women's chances of experiencing a trauma over men's and, given traumatic exposure, their vulnerability to develop PTSD.

PTSD illustrates how the *DSM* has come to exert tremendous influence over cultural conceptions of mental illness. Largely as a result of the changes in the *DSM-III-R* and, especially, the *DSM-IV*, PTSD became psychiatry's emblematic condition. The term "PTSD" pervades both media discussions and lay descriptions of responses to trauma.[69] Articles about PTSD in the medical literature doubled between 1985 and 1995 and then doubled again between 1995 and 2005. By 1999 more than 16,000 publications addressed PTSD.[70] A large trauma industry became widely institutionalized in schools, hospitals, police and fire departments, government, and industry as well as in mental health facilities.[71] A new profession of grief and trauma counselors flocked to the scenes of catastrophes. The recent centrality of PTSD is even more remarkable because it is the rare major *DSM* diagnosis that the drug industry, which has yet to develop any effective product to alleviate traumatic symptoms, has not promoted.

Feminists thus influenced the *DSM-III-R* and *IV* in major ways. One prong of feminist activism fought diagnostic expansion of conditions that would stigmatize growing numbers of women. Another prong vigorously propounded, first, the emergence of the MPD diagnosis and then enlarging the PTSD diagnosis to include far more victimized women. Although MPD has almost completely disappeared, "PTSD is perhaps the fastest growing and most influential diagnosis in American psychiatry," historians Paul Lerner and Mark Micale observe.[72]

DSM Diagnoses and Drugs

Before the early 1970s, most drug marketing did not target particular diagnoses. Instead, the minor tranquilizers were touted for their ability to relieve such broad states as "tension," "stress," or "anxiety." The 1971 FDA ruling discussed in chapter 3 changed this situation and forced pharmaceutical companies to market their products as remedies for specific diseases. The *DSM-III* provided a capacious range of conditions that could meet the FDA specificity mandate. By the time the *DSM-IV* was published in 1994, the drug industry had become the most potent force propelling psychiatric diagnoses into cultural consciousness.

Panic Disorder and Xanax

In 1981 the benzodiazepine Xanax (Upjohn's trade name for alprazolam) became the first new psychotropic drug to enter the market after the publication of the *DSM-III*. It was the initial test of how drug companies could use the new diagnostic system to meet the FDA mandate for specificity. Upjohn hoped to promote Xanax as a more advanced version of Valium, which was scheduled to lose its patent in 1985 and, in any case, had been discredited during the anti-tranquilizer movement in the 1970s. The drug had some of the antianxiety properties of the benzodiazepines but also some antidepressant aspects.[73]

Upjohn originally intended to market Xanax as an antidepressant, but without a study demonstrating its effectiveness with inpatients the FDA blocked it.[74] Upjohn shrewdly shifted its target and promoted Xanax as a remedy for panic disorder, an anxious condition that diagnostic manuals before the *DSM-III* had not specifically recognized. Backed by an aggressive advertising campaign, within a few years Xanax became the top-selling psychotropic drug and is still widely prescribed.[75]

Perhaps most important, Xanax showed the drug industry how valuable the *DSM* diagnoses could be for marketing their products. It was the first of many successful attempts to commercialize the manual's conditions. The idea of a tranquilizer that worked across a spectrum of nervous states was dead. "Henceforth," Edward Shorter observes, "magic bullets would match disease labels: There would be only anxiolytics for anxiety, antidepressants for depression, and antipsychotics for what everybody was calling 'schizophrenia.'"[76]

Major Depression and the Selective Serotonin Reuptake Inhibitors

For a century after George Beard coined the label "neurasthenia" in 1869, the most common psychic complaints among medical and psychiatric outpatients involved a mélange of anxious, depressive, and psychosomatic symptoms. The minor tranquilizers continued to target such undifferentiated states of "nerves," "nervous breakdowns," or "fatigue" through the early 1970s. In contrast, melancholic depression was a relatively uncommon but extremely severe state often associated with periods of hospitalization. It was typically treated with antidepressant drugs, such as the MAOIs and imipramine, that were not used for other conditions. General physicians widely prescribed tranquilizers but not antidepressants, which remained in the sphere of psychiatrists. Accordingly, surveys in the 1960s and early 1970s indicated that about 10 times more prescriptions were written for tranquilizers than for antidepressants.[77]

The specific nature of the *DSM-III* diagnoses created a vacuum: What label could physicians apply to the diffuse complaints that so many of their patients presented and sought medication for? Fortunately for both parties, the *DSM-III* solved the problem of what diagnosis best fit an unspecific, extremely broad condition: major depressive disorder.[78] As the previous chapter noted, the *DSM-III's* formulation of MDD conflated commonly occurring but relatively mild depressive symptoms of the neurasthenic package with rare yet severe melancholic symptoms such as extreme slowing of thought and movement and nonresponsiveness to external events.[79] Eminent diagnostician Donald Klein recognized the incoherence of MDD and urged Spitzer to separate these thoroughly distinct conditions in the *DSM-III-R*: "I think that the distinction between the relatively autonomous depression and the relatively reactive depression is a strikingly important one that should be present in this revision."[80] Spitzer ignored this suggestion. The result, psychiatrist Herman van Praag wrote in 1990, was that "today's depression classification is as confusing as it used to be 30 years ago. All things considered, the present situation is worse. Then, psychiatrists were at least aware that diagnostic chaos reigned and many of them had no high opinion of diagnosis anyhow. Now, the chaos is codified and thus much more hidden."[81]

The confounding of one condition that was very common but usually not severe with another that was infrequent but very serious transformed diagnostic practices. While depressive diagnoses were already increasing during the 1970s, their major expansion arose after the *DSM-III* was published in 1980. From that time onward, rates of MDD surged. Between 1987 and 1997, the proportion of the US population receiving outpatient treatment for conditions called "depression" increased by more than 300 percent. While in 1987, 20 percent of psychiatric outpatients had a diagnosis of some kind of mood disorder, most of which were MDD, 10 years later depressive diagnoses nearly doubled, to account for 39 percent of all outpatient conditions.[82] This explosion occurred despite the fact that, in contrast to PTSD, the MDD diagnostic criteria were virtually unaltered from the *DSM-III* to the *DSM-III-R* and *DSM-IV*. The use of the criteria, not the criteria themselves, changed over time.

The swell of depression diagnoses is especially noteworthy when contrasted with its neurasthenic sibling, anxiety. Rates of any anxiety diagnosis for treated patients rose much more slowly than those of depression, from 10.5 percent in 1987 to 12.5 percent in 1997. By the latter year diagnoses of mood disorders were more than three times as common as anxiety diagnoses in office-based psychiatry.[83]

More recent figures present a mirror image of the overwhelming dominance of anxiety in general medicine and psychiatry during the 1950s and 1960s. In 2002, 51.7 million outpatient visits were made for mental health care. Depression accounted for 21 million of these, compared to just 6.2 million for anxiety.[84] Likewise, by the early part of the twenty-first century, general physicians were more than twice as likely to make diagnoses of depression as of anxiety.[85] Among office-based physicians, depression accounted for 42 percent of mental health diagnoses from 2007 to 2010.[86] For whatever actual problems people sought mental health care, the treatment system and, in all likelihood, patients themselves were calling them "depression." For example, depression is the single most common topic of online searches for pharmaceutical and medical products, attracting nearly 3 million unique visitors over a three-month period in 2006.[87]

Although drug companies played no direct role in constituting the DSM diagnoses, they were probably the most important influence over how the manual's conditions entered public awareness. Comparable to their predecessors, the benzodiazepines, the SSRIs that came on the market in 1987 were anything but specific to any particular type of mental disorder. Instead, like Miltown, Librium, and Valium, they acted on the broad array of conditions that the neurasthenic tradition encompassed. They increase levels of serotonin in the brain, which in turn influence a wide range of moods and feelings rather than any particular condition. By the late 1980s, however, their manufacturers had no choice but to market them as remedies for some specific mental disorder. The pharmaceutical industry's exploitation of the DSM categories was a key factor transforming anxiety diagnoses into depressive ones.

MDD proved to be the perfect diagnostic companion for the new SSRIs. For drug marketers, the beauty of the diagnosis was that it packaged a label, "major depression," which was associated with a very serious mental disorder but also could be diagnosed through extremely common symptoms such as sadness, fatigue, and sleep and appetite problems. Moreover, the MDD criteria required symptoms to persist for just two weeks. Drug companies realized that marketing the SSRIs as "antidepressants" had the potential both to overcome the stigma that antianxiety drugs had acquired in the 1970s and to appeal to an extremely large market. The pharmaceutical industry could not have predicted, however, the extent of the financial bonanza that the MDD diagnosis would reap for them. In 1987, the first year it entered the market, Prozac brought in $125 million. This amount nearly tripled to $350 million in the following year. By the beginning of the new century, worldwide sales had reached $2.6 billion each year.[88] "It would be hard to overstate," writes historian David Herzberg, "the degree to which Prozac became famous as a commercial product."[89]

When the SSRIs were approved in the late 1980s, antianxiety drugs were about twice as likely to be prescribed in outpatient visits as were antidepressants.[90] At that point, this proportion changed abruptly. Between 1985 and 1993/1994, prescriptions for antianxiety drugs plunged from 52 to 33 percent of all psychopharmacological visits, and the number of users of antianxiety drugs grew very slowly after that, rising from 5.5 million to 6.4 million in 2001.[91] Conversely, from 1996 to 2001, the number of SSRI users increased rapidly, from 7.9 million to 15.4 million. By 2000, the antidepressants were the best-selling category of drugs of any sort in the United States; fully 10 percent of the US population was using an antidepressant.[92] These drugs were used so widely in general medical practice that in 2003/2004, 310 of every 1,000 female patients received a prescription for an antidepressant. The use of the SSRIs continued to grow, and by 2006 Americans had received more than 227 million antidepressant prescriptions, an increase of more than 30 million since 2002.[93] Antidepressants were prescribed for mood, anxiety, and many other conditions, gaining unchallenged control of the market the anxiolytic drugs once held.

The next major development occurred in 1997, when the FDA eased its rules regulating how drug companies could advertise their products directly to consumers. Spending on advertising immediately increased from $595 million to $843 million and by 2000 had exploded to $2.5 billion.[94] Direct-to-consumer ads, accompanied by elaborate websites, self-diagnostic tests on the internet, and slick media campaigns, propelled the *DSM* diagnoses into the public spotlight. These ads contextualized the purely descriptive MDD criteria so that normal responses to stressors could be seen as indications for taking medication.

An ad for the SSRI Paxil was emblematic (figure 5.1). Like the promotions for the benzodiazepines, it portrayed an extremely common psychosocial problem, the alienation of a woman from her husband and child. Yet, in contrast to the earlier depictions, depressive symptoms were the *cause* of what stood between the woman and realizing her life, not the consequence of, say, her spouse's affair or her child's behavior problems. Moreover, the ad used the most common symptoms of the MDD

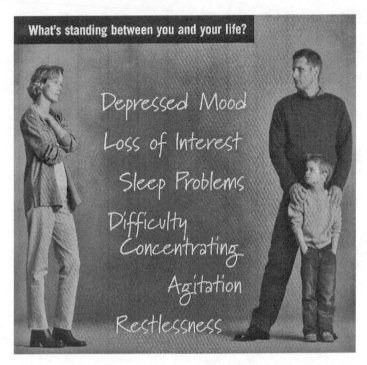

Figure 5.1. Advertisements for the SSRIs typically presented symptoms as the causes, not results, of psychosocial problems. *Cosmopolitan*, February 2003.

diagnosis—depressed mood, loss of interest, sleep problems, difficulty concentrating, agitation, and restlessness—omitting any mention of the melancholic symptoms of worthlessness, inappropriate guilt, or recurrent thoughts of death.

The symptoms of MDD, not the family situation, created the woman's unhappiness. It follows that taking Paxil to ease the symptoms will also solve her relational problems. Indeed, the bottom panel (not shown) indicates that after taking Paxil the woman was happily reunited with her son (although not, perhaps tellingly, her husband). The ad brilliantly used the *DSM* diagnostic criteria to sell a product that resolves social difficulties and restores personal happiness. Advertising campaigns such as Paxil's were extraordinarily successful in capitalizing on the MDD diagnosis.

Another aspect of the promotional campaigns for the SSRIs was their successful invocation of the *DSM* notion that depression is a disease that is external to the person who experiences it. Ads for earlier classes of drugs showed symptoms as aspects

Figure 5.2. Drug advertisements before the *DSM-III* usually portrayed symptoms as results, not causes, of psychosocial problems. *JAMA*, January 6, 1975.

of individuals and their life experiences. An ad for the antidepressant Triavil, for example, explicitly claimed that it is the *result* of situations such as the "empty nest syndrome" (figure 5.2). The Paxil ad, in contrast, portrayed symptoms that exist independently of the woman. Such depictions influenced people to view depression (and other mental disorders) as a self-contained disease, not as a problem in living. As Schnittker's Google nGram analysis showed, the use of the term "have depression" skyrocketed after 1980 and surpassed mentions of "am depressed" by 2000.[95]

Ironically, the SSRIs are probably least effective for the condition—depression—that they are marketed for. They are less successful in treating melancholic depression than older medications and have marginal impacts on reactive depression, but they are more effective with anxious conditions.[96] Despite this evidence, the need to use the *DSM* straitjacket led to their initial promotion as "antidepressants." Whatever the SSRIs do has little relationship to any specific *DSM* diagnosis but cuts across many diverse syndromes.

One of the most interesting aspects of recent diagnostic history is that drug advertising in the 1950s and 1960s, which touted the tranquilizers as remedies for a broad range of problems in living, was more accurate than later efforts that promote their current counterparts as treatments for specific diseases. The regulatory framework that the FDA imposes, however, bans the former and mandates the latter, leaving drug companies with no choice but to misleadingly use the *DSM* diagnoses.

SOCIAL PHOBIA AND PAXIL

By the late 1990s the SSRIs were imminently scheduled to lose their patents for treating depression. Therefore, drug companies had to reposition their use for other specific *DSM* mental disorders.[97] This involved, first, conducting trials with the relevant groups; second, obtaining FDA approval for a new use; and, finally and most important, creating a new diagnostic market for the drug. Social phobia (later called "social anxiety disorder") and Paxil provides the best example of this process.

As the neurasthenic package implied and as epidemiological research indicated, most depressed people are also anxious and most anxious people are also depressed. This creates a potential market for products that focus on the anxious part of the broad range of conditions that the SSRIs can treat. Neither the *DSM-III* nor *III-R*, however, had a category of mixed anxiety-depression.[98] In the 1990s, social phobia became the major target of the SSRIs for the anxious component of the neurasthenic tradition.

Most of the conditions that appeared in the *DSM-III* were not new but contained far more specific descriptions than the general conceptions found in previous manuals. Social phobia exemplified this type of diagnosis. It featured extreme anxiety over situations in which people are exposed to the scrutiny of others, such as when they must speak in public, have their performance appraised, or attend social events that involve interacting with strangers. Their distress often leads to avoidance of these situations.[99]

Psychiatric manuals did not mention this condition until 1980 when the *DSM-III* provided brief diagnostic criteria for it.[100] They attempted to avoid the implication that ordinary shyness and avoidance of uncomfortable social situations were mental disorders: "Avoidance of certain social situations that are normally a course of some distress, which is common in many individuals with 'normal' fear of public speaking, does not justify a diagnosis of Social Phobia."[101] In addition, social phobia was a residual condition that was not diagnosed in the presence of other mental disorders such as MDD or avoidant personality disorder. Because most people who met diagnostic criteria for a social phobia also met these other criteria, it was no surprise that, when it first appeared, the manual noted, "The disorder is apparently relatively rare."[102] Initial studies in the early 1980s that used the *DSM-III* criteria indicated that about 1–2 percent of the population reported social phobias.[103]

The *DSM-III-R* broadened the earlier criteria. It eliminated the requirement that the fear be "irrational" and involve "compelling desire to avoid" the situation and allowed diagnosis in the presence of any other *DSM* condition. It also provided examples that were very common, including "being unable to continue talking while speaking in public . . . or not being able to answer questions in social situations."[104] The *DSM-IV* did not make any major changes to the more expansive *DSM-III-R* criteria, although it added an alternative label, social anxiety disorder (SAD).[105] A major community study using the *DSM-III-R* criteria indicated that over 13 percent, or one of eight people, had a social phobia at some point in their lives.[106] Other surveys indicated that up to 20 percent of the population suffered from SAD.[107] Indeed, by the early 2000s, social phobia was one of the two most common mental disorders.[108] How did the number of people with this condition more than *quintuple* over such a short time?

Rising rates of social phobia stemmed from seemingly small changes in how community surveys use the *DSM* criteria. The *DSM-III-R* and *IV* alterations were important because the boundaries between mental disorders and extremely common but normal anxiousness over situations like public speaking are quite porous. For example, having an unreasonably strong fear of public speaking is the most common symptom of social phobia. Changing just the wording of a question that asked about having extreme distress when "speaking in front of a group you know" to "speaking in front of a group" doubled the number of positive responses. Likewise, changing the criteria from have "a compelling desire to avoid" fear-inducing situations to having "marked distress" in these situations resulted in a sharp increase in the reported amount of social phobia.[109]

The pharmaceutical industry quickly recognized the potential of a new condition that was hard to distinguish from very common anxiety-provoking situations and ordinary shyness. "Every marketer's dream," Paxil's product director exclaimed, "is to find an unidentified or unknown market and develop it. That's what we were able to do with social anxiety disorder."[110] In 1999 the FDA approved Paxil specifically for the treatment of SAD. Its manufacturer, GlaxoSmithKline, mounted a huge advertising campaign, whose key message was that "Paxil is the first and only FDA-approved medication for the treatment of social anxiety disorder."[111] The ad capitalized on the epidemiological findings, noting that more than 10 million Americans suffer from SAD. GlaxoSmithKline spent more than

$90 million dollars on a barrage of print and television ads that urged viewers to "imagine being allergic to people" (figure 5.3).

These ads were just the tip of the iceberg of a gigantic public relations campaign by the firm Cohn & Wolfe that aimed to fundamentally reshape public perceptions of social anxiety from being shy and uneasy in social situations to having a mental disorder treatable with a drug.[112] Other aspects of the promotion involved placing stories in the news media, often using celebrity figures, psychiatric experts, and testimonies from members of consumer advocacy groups about the pervasiveness

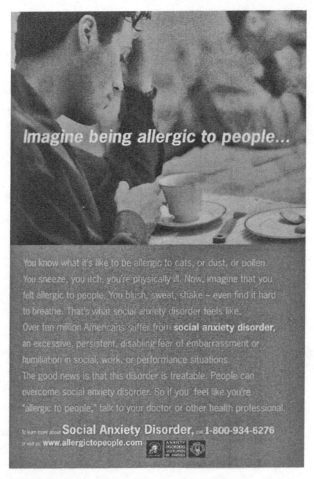

Figure 5.3. Some advertisements raised awareness of a new disorder, such as social anxiety disorder, rather than push a particular drug. Cohn and Wolfe, 1999.

of SAD. The immense campaign was a huge success: Paxil became the largest-selling antidepressant at the time, to the tune of $3 billion a year. Social anxiety disorder became a common and well-recognized mental illness. Indeed, *Psychology Today* named it "the disorder of the decade" of the 1990s.[113]

The creation of social anxiety disorder and the accompanying need to medicate it shows how drug companies can exploit the *DSM* criteria and turn distressing but normal emotions such as shyness into widespread mental disorders. Formerly shy people can now see themselves as having an illness with a pharmaceutical solution. Allen Frances later expressed regret that the *DSM-IV* did not increase the diagnostic threshold for social phobia: "We clearly had our collective heads in the sand and badly miscalculated."[114]

BIPOLAR II AND THE NEW ANTIPSYCHOTICS

The emergence of social anxiety disorder brought a part of the neurasthenic package that the MDD diagnosis did not contain into the range of conditions that the SSRIs treated. A second expansion of the diagnostic objects of psychotropic drugs capitalized on another new condition in the *DSM-IV*, bipolar II. As with SAD, marketing campaigns created a huge demand for this diagnosis and accompanying antipsychotic medication.

The relationship of bipolar disorder (or manic depression) to depression had been a central aspect of discussions about psychiatric diagnosis since Kraepelin's writings. The German diagnostician maintained that all severe depressions, whether unipolar or bipolar, were part of a single entity that he called "manic depression," which had a common fluctuating course and outcome. He contrasted this condition with dementia praecox, or schizophrenia, which had progressively deteriorating symptoms. Unlike Kraepelin, the *DSM-I* and *II* contained separate categories of depressive reactions and manic-depressive reactions within the broader class of psychotic disorders.

Another line of thought also cast doubt on the Kraepelinian conflation of unipolar and bipolar conditions. The Washington University group's research indicated that patients with depression versus manic depression had distinct family backgrounds.[115] The Feighner criteria reflected these findings: their category of primary affective disorders had separate subcategories for depression and mania but did not mention bipolar conditions. For the most part, the *DSM-III* reflected the Washington University classification. It changed the name "manic depression" to "bipolar disorder" and placed it within a distinct affective disorders class. Its bipolar diagnosis contained separate codes for manic and depressed phases.[116] In addition, as the previous chapter explored, it thoroughly intertwined psychotic and neurotic depressions within the MDD diagnosis. The effect of the *DSM-III* classification was to draw clear lines between bipolar conditions and both depression and schizophrenia.[117]

The *DSM-III-R* did not make major changes in the bipolar criteria. However, the *DSM-IV*, contradicting its principle of not incorporating new diagnoses that did not have overwhelming empirical support, included a novel condition of bipolar II disorder. This diagnosis thoroughly blurred the sharp boundary between bipolar and depressive conditions that the *DSM-III* had established. Bipolar II was unique in psychiatric history, requiring the presence of MDD and a hypomanic episode that needed to persist for four *days*. This brief period "involved no psychotic features" but required three from among seven symptoms including inflated self-esteem, decreased sleep, talkativeness, and distractibility.[118] Jerome Wakefield notes how bipolar II can lead to many false positive diagnoses:

> For example, a period of intense romantic involvement may satisfy the criteria for hypomanic episode if it lasts at least four days and the lover experiences an elevated, expansive mood during which he or she is three or more of the following: sexually indiscreet, distractible, physically active, talkative, high in self-esteem, lacking sleep, and having thoughts racing. If the lover is spurned and goes into a depressed tailspin for two weeks that satisfies the criteria for major depressive episode, then he or she will qualify for a diagnosis of bipolar II disorder, even if no disorder is really present.[119]

Because it was not difficult to meet both the MDD and the hypomanic criteria, the floodgates opened for the first epidemic of a condition with a label that had been traditionally been associated with a psychosis.

Community studies based on the *DSM-III* criteria had estimated the lifetime prevalence of bipolar conditions at 1–2 percent of the population. Studies using the *DSM-IV* criteria indicated that far more people were bipolar *in any single year* than previous lifetime estimates: "Bipolar disorder affects approximately 5.7 million adult Americans, or about 2.6% of the U.S. population age 18 and older every year." Other research found that the lifetime risk of bipolar I and bipolar II to be as high as 10 percent of the population![120]

Before the *DSM-IV* it would have been unthinkable for a drug company to promote an antipsychotic for a mass market. Yet, the year after this manual came out, Abbott Laboratories touted the mood stabilizer Depakote, which had been used to treat epilepsy, as a treatment for manic symptoms. The next year the FDA approved the atypical antipsychotic Zyprexa for acute manic episodes. In the following decade, the FDA approved Abilify and Seroquel as treatments for bipolar conditions and as add-ons for the treatment of MDD. AstraZeneca even set up a website, isitreallydepression.com, in a bid to capture the lucrative depression market for Seroquel (figure 5.4).

Figure 5.4. Advertisements for antipsychotics used to treat bipolar disorders tried to capture the huge antidepressant market, asking, "Is it really depression?" *Huffington Post Contributor Platform,* January 31, 2012.

The drug industry had hit another jackpot. Zyprexa's producer, Eli Lilly, launched a huge marketing campaign. Echoing the advertising campaigns of the 1950s and 1960s, Zyprexa was touted as a solution for marital problems stemming from a spouse's rages and unpredictable behaviors. The ads presented signs such as anger, elevated mood, irritability, or agitation as possible indicators of a bipolar condition. Bipolar disorders not only generated massive profits but also attracted broader public attention. Psychoses, which had always been considered rare conditions, were suddenly found everywhere. Psychologist Kay Jamison wrote a best-selling memoir, *An Unquiet Mind*, in 1995 as well as another book, *Touched by Fire*, which celebrated the creativity and contributions that many artists and authors with bipolar conditions had made. A cover story in *People* explored actress Catherine Zeta-Jones's battle with bipolar disorder. One website about bipolar conditions has a "narrowed" list of "nine must see movies about bipolar disorder" produced since 1993.[121]

The diagnostic changes in the *DSM-IV* allowed drug companies to propel a formerly rare condition to prominence as a widely celebrated cultural phenomenon. Bipolar II exemplifies, as historian Andrea Tone observes, a diagnosis that captures "the relentless expansion of illnesses to accommodate new medications that purport to treat them."[122]

Childhood Bipolar Disorder and Mood Stabilizers

The emergence of bipolar II was not the only major change involving bipolar conditions. In the 1980s a Harvard child psychiatrist, Joseph Biederman, proposed that many young children who were given labels of attention deficit/hyperactivity disorder actually had bipolar disorders. Biederman's contention was unprecedented in psychiatric history. Previously, bipolar conditions were thought to emerge in the late teenage or early adult years. To the extent that they were specifically age linked, they were traditionally connected to midlife. The first psychotic condition listed in the *DSM-I* was involutional psychotic reaction, which definitionally emerged among middle-aged men and, especially, women.[123] The *DSM-II* maintained a category of involutional melancholia that arose in middle age but did not mention any psychosis among the childhood disorders it listed.[124] The *DSM-III* vastly expanded the category of disorders usually first evident in infancy, childhood, or adolescence, but this extensive group did not contain any condition resembling a bipolar psychosis.[125]

The children that Biederman saw demonstrated temper tantrums, chronic irritability, constant distraction, and many other oppositional behaviors from a very early age. He proposed the striking thesis that, although they didn't have any manic episodes, they actually had childhood bipolar disorders (CBD) that should be treated with mood stabilizers, a type of antipsychotic drug, rather than stimulants. The best-selling *The Bipolar Child*, by Demitri and Janice Papolos, propelled CBD into the culture. Programs such as *Oprah* and *20/20* hailed the book, generating a large following for the authors among parents who had given up hope for dealing with their extremely troublesome children. Numerous websites and self-help groups emerged devoted to this condition.[126] In August 2002 the cover of *Time* magazine featured the story "Young and Bipolar."

Pharmaceutical companies were immediately attracted to a condition that began in childhood and often required continuous and lifelong antipsychotic drugs. The widespread media attention to pediatric bipolar conditions led many parents to seek medication that would presumably help sedate their out-of-control children. Although none of the *DSMs* contained a specific diagnosis of CBD, their bipolar criteria also did not prevent diagnosing children with this condition. The new antipsychotics were not explicitly approved for CBD, but they were nonetheless extensively prescribed for it.[127]

The result was a remarkable growth in bipolar diagnoses among children who made office visits for mental health problems. In 2007 a national survey discovered an astonishing *fortyfold* increase in the number of children and adolescents treated for bipolar disorder from 1994 to 2003. Rates of this condition leaped from less than 0.5 percent in 1994 to 6.67 percent in 2003. By the latter year more than 1.6 million children and adolescents received bipolar diagnoses in private office visits. This figure is all the more remarkable because such diagnoses were virtually nonexistent just 10 years before.[128]

The spectacular rise in rates of CBD provides perhaps the most dramatic instance of the discovery of a new mental disorder. The common denominator among youth treated for this condition seems to be that their conduct is extremely disturbing to their parents or teachers. Pediatric bipolar diagnoses and resulting prescriptions for antipsychotics provide ways to pacify disruptive behavior. Many experts question the existence of the disorder at all and maintain that it actually reflects the petulant, angry, and ungrateful behaviors that mark normal children and adolescents.[129] The fact that Biederman, its leading promoter, had received millions of dollars in consulting fees from the drug industry fueled skepticism over the legitimacy of the childhood bipolar diagnosis and ensuing medication treatments.[130] "The CBD fad is the most shameful episode in my forty-five years of observing psychiatry," *DSM-IV* chair Allen Frances admitted.[131]

Expanding Normality: Gender and Sexuality

Most changes in *DSM-III-R* and *DSM-IV* criteria increased the range of pathology. Diagnoses related to sexual orientation are the major exception to this trend. The furor over homosexuality in the early 1970s had contributed both to the downfall of the *DSM-II* system and to the discrediting of psychoanalysts, the most fervent defenders of the diagnosis. After a number of false starts, Spitzer created a category of ego-dystonic homosexuality (EDH) in the *DSM-III* that applied only to people who were dissatisfied with their homosexual orientation.

A number of psychiatrists and psychologists objected to the EDH diagnosis. First, they claimed it was inconsistent with the *DSM*'s principle that distress in itself was not the basis for disorder status. If homosexuality was a normal variant of sexuality, as the *DSM* accepted in 1973, then distress over it could not be pathological. Second, critics asserted that it was inconsistent for the EDH category to apply to people disturbed by their homosexuality but not to those who were disturbed by their heterosexuality.[132]

In response, the *DSM-III-R* deleted the EDH diagnosis but added a category of sexual disorder not otherwise specified (SDNOS) for "persistent and marked distress about one's sexual orientation."[133] This incoherent diagnosis maintained a state where distress over a condition constituted a mental disorder even though the condition itself was not disordered. In a nod to the complaints that the former EDH diagnosis singled out gay people, the new condition made "distress about one's sexual orientation" the generating force, although the entire recorded psychiatric literature up to that time contained just a single case of ego-dystonic heterosexuality.[134] The APA Board of Trustees approved the deletion of EDH and the movement of distress over sexual orientation to the SDNOS category.

The contention over homosexuality ended with a whimper when the *DSM-IV* eliminated any mention of sexual orientation, which could no longer generate a diagnosis of mental illness. The whole saga was a rare case where the realm of mental disorder in the *DSM* contracted rather than expanded.

A Successful Revolution?

The two decades following the introduction of the *DSM-III* restored the high status the psychiatric profession had attained during the postwar decades. After 1980, the *DSM* diagnoses reaped high rewards for many interests. Most notably, the classification rescued a pharmaceutical industry that had been reeling from the attacks on the tranquilizing drugs in the 1970s. The numerous diagnostic targets the new manual presented were godsends for drug companies. In turn, psychiatry departments, which had been considered medical backwaters, flourished as drug industry money flowed to their members who researched the effect of psychotropics on the various new diagnostic categories. The NIMH, once mired in political controversies, was transformed into an agency that sponsored investigations into genuine medical disorders and their biological treatments. The huge prevalence of the major *DSM* categories that epidemiological studies uncovered provided psychiatric researchers with the justification of confronting a variety of "public health epidemics." Mental health advocates, too, gained much broader potential support from the many millions of persons who presumably suffered from the *DSM* ailments. Clinicians, who initially fought the imposition of the *DSM-III* classification, rapidly came to support diagnoses that garnered them third-party reimbursements and elevated their status as healers of diseases. Not least, patients gained names for their afflictions, compensation for their visits to physicians and other mental health professionals, and drugs to alleviate their distressing conditions.

The developers of the *DSM-IV* expected that this manual would be another step in the continuous progress of psychiatric research: "The highest purpose of the *DSM-IV* is that it encourage and facilitate the research that will render it obsolete."[135] Around the turn of the century, however, cracks in the *DSM* model arose from a highly unlikely source: psychiatric researchers. This group, which had been the primary force behind the *DSM-III, III-R,* and *IV*, constituted the major critics of the diagnostic system. Whatever advantages the *DSM*'s descriptive categories provided for many interest groups, by the early twenty-first century the manual's various diagnoses had become millstones for research purposes. "At present, psychiatry is in a state of flux," two eminent diagnosticians wrote in 2003, "and advances in neuroscience and genetics are soon likely to challenge many of its current theoretical underpinnings, particularly those related to the causation and definition of mental disorders."[136] The *DSM-III-R* and *IV* revisions mainly involved tinkering with the *DSM-III* diagnoses and adding some new categories. In contrast, researchers envisioned that the next edition would radically transform the *DSM*. Their efforts to emulate the successful *DSM-III* revolution would have a very different outcome in the *DSM-5*.

six | The *DSM-5*'s Failed Revolution

The *DSM-III* not only transformed the process of psychiatric diagnosis but also thrust the manual into cultural prominence. The inauguration of new diagnoses or, more rarely, the elimination of current ones in the *DSM-III-R* and *IV* maintained the manual's high profile. Yet the attention paid to these revisions paled in comparison to the outcries that accompanied the emergence of the next version of the diagnostic manual, the *DSM-5*.[1] Earlier clashes had involved specific topics such as the status of homosexuality as a mental disorder, the elimination of the term "neurosis," the iatrogenic aspect of multiple personality disorder, or the stigmatizing potential of a PMS diagnosis. They rarely involved ad hominem arguments. The run-up to the *DSM-5*, in contrast, centered on issues concerning the essential qualities of psychiatric diagnoses and the motives of the major players involved in the process. The spectacle played out in the public arena: in the five years before the manual was issued in 2013, nearly 3,000 print and online articles discussed its development.[2]

The framing of the *DSM-5* revision was distinctly different from the construction of the previous two manuals. The developers of the *DSM-III-R* and *IV* portrayed them as steps in the steady improvement of the *DSM-III*. They strove to build upon the 1980 manual and never considered replacing its descriptive, categorical, and theory-neutral structure. The participants in these revisions assumed that constant changes in diagnostic criteria would create a difficult learning curve for clinicians, make it impossible for researchers to accumulate standardized knowledge, and force mental health institutions to make massive changes in administrative procedures. In contrast, the leaders of the *DSM-5* process set out to overturn the core principles of the diagnostic system. Comparable to the *DSM-III* upheaval, they intended to make "a paradigm shift" that would revolutionize psychiatric diagnoses.[3]

The attempt to implement a new diagnostic system in the *DSM-5* was just one of the central differences from the previous two revisions. The major focus of the process shifted from the relatively technical concern with reliability to the more basic consideration of validity: Do the *DSM* diagnoses accurately portray the basic aspects of mental disorders? Achieving validity required some understanding of the causes of mental disorders, which the extant *DSMs* had not yet provided. In addition, the latest revision resurrected the conflict between clinicians and researchers that was at the heart of the *DSM-III* process but that had little influence on the following two manuals. The development of the *DSM-5* was also far more decentralized than the highly controlled processes of previous revisions had been. The sweeping autonomy the various diagnostic work groups possessed led to much internal disagreement and disorganization.[4] Another difference was that the close cooperation between the American Psychiatric Association and the National Institute of Mental Health that marked the design of all previous *DSMs* disintegrated; the *DSM-5* saga resulted in a profound rupture between the professional organization and its former federal ally. Finally, the development of the *DSM-5* took place within a scientific culture that embraced a brain-based view of mental disorders, which made the *DSM*'s principle of theory neutrality difficult to sustain.

Disturbing Findings from Neuroscientific Research

Theory neutrality was a core principle of the *DSM-III, III-R*, and *IV*. Each manual relied on descriptive diagnostic criteria and decision rules that entailed no etiological assumptions. Therefore, psychiatrists and other mental health professionals of all theoretical persuasions could use them. Since 1980, however, its diagnostic manual notwithstanding, psychiatry had undergone a massive transformation. In practice, the field had replaced the pluralist combination of psychotherapy, psychosocial interventions, and drug treatments that characterized it during the postwar period with a "bio-bio-bio" model that emphasized brains, genes, and medications.[5] These internal changes reflected the movement of scientific culture more generally.

In the final decades of the twentieth century, the field of neuroscience began to explode; membership in the area's professional society increased from just

500 in 1969 to 6,351 in 1979 and to 13,433 in 1989.[6] The spur for this growth was the discovery of neuroimaging techniques such as computed tomography (CT) scans and functional magnetic resonance imaging (fMRI), which allowed scientists previously unthinkable opportunities to visualize and measure brain activities and molecular processes. President George H. W. Bush declared the 1990s "the decade of the brain." By 2005 the National Institutes of Health was providing about $4 billion each year to support neuroscientific studies. In what is considered to be one of the major achievements in the history of science, the human genome was deciphered in 2003. Further, the development of genome-wide association studies in the first decade of the twenty-first century provided researchers the opportunity to study hundreds of thousands or even millions of alleles that might reveal possible genetic abnormalities underlying mental disorders. Publications on the brain grew from 13,000 in 1970 to more than 60,000 each year since 2010.[7]

By the end of the twentieth century, there seemed to be no doubt that the cutting edge of research on mental disorders lay in "biochemical abnormalities, neuroendocrine abnormalities, structural brain abnormalities, and genetic abnormalities."[8] Yet neither the *DSM-III* nor its revisions incorporated any of these innovations. It seemed inevitable that the manual's theory-neutral approach would have to yield to a classification that better reflected the brain-based approach to mental disorders.

There was, however, a large fly in the ointment. The findings from neuroscientific research were far from what researchers expected. They revealed a striking disconnect between the *DSM* model and their understanding of mental disorders.[9] Perhaps the most important conclusion emerging from genetic studies was that, contrary to the *DSM* assumption of disorder specificity, genes for virtually all psychiatric disorders are nonspecific. No disorder corresponds to a distinct gene or group of genes; instead, all share large amounts of genetic vulnerability with other conditions: any genetic variant that is tied to one diagnosis is also associated with multiple others.[10] In addition, the most characteristic symptoms of mental disorders were widely distributed across diagnoses and not localized within any particular one. The result was a striking lack of progress in the understanding of mental disorders. For example, in regard to schizophrenia, one group of researchers concluded in 2003, "In some ways we are not much further ahead than Kraepelin; diagnosis is still based on the same clinical observations."[11]

A review of 537 studies with 21,427 participants that compared fMRI images of schizophrenia, bipolar disorder, major depressive disorder, anxiety disorders, and obsessive-compulsive disorder with non-disordered controls provides a recent example. It found many differences between the diagnosed groups and the controls but none that were linked to any particular diagnosis.[12] Family studies, too, indicated the nonspecificity of mental disorders. One instance was families with a depressed member, which also have elevated rates of bipolar disorder, drug addiction, alcohol abuse, and eating disorders, among others.[13] Epidemiological research buttressed these findings: over two-thirds of depressed people suffer from some anxiety disorder, and over three-quarters also have some other mental illness.[14] Likewise, the major forms of anxiety disorders are not distinct but reflect the "activation of one and the same underlying anxiety response."[15] The various personality disorders also show a huge overlap with each other. People with one personality disorder are the exception; the rule is that they qualify for multiple diagnoses.[16] In other words, the major *DSM* conditions were not discrete but highly overlapping.

Research also produced another key finding at variance with *DSM* assumptions: a small number of general vulnerabilities rather than a host of particular causes best characterized mental disorders. For example, psychologist Robert Krueger examined reports of symptoms in a large national study without regard to particular categorical disorders. He found that all the symptoms of anxiety and mood disorders were aspects of a single broad condition, which he suggested might best be called "internalizing." These conditions contrasted with a second set of equally broad "externalizing" conditions such as ADHD, alcohol abuse, drug dependence, antisocial personality disorder, and conduct disorder.[17]

Biological research even called into question Kraepelin's time-honored distinction between the symptoms, courses, and outcomes of schizophrenia and bipolar disorder.[18] Genome-wide association studies, in contrast, demonstrated that these conditions share about two-thirds of their genetic risk.[19] Most patients have indications that typify both schizophrenic and bipolar states; moreover, as they move through the life course, their dominant symptoms shift across these disorders. These presumably separate diagnoses have high genetic overlap with each other as well as many genes in common with other conditions such as major depression, ADHD, and autism: "[The] boundaries between schizophrenia and other psychiatric disorders are indistinct."[20] Some researchers go even further, claiming that a single dimension, similar to the g factor in intelligence that provides a summary measure of general mental ability, accounts for all types of psychopathology across the life course. "Today's patient with schizophrenia was yesterday's boy with conduct disorder or girl with social phobia (and tomorrow's elderly person with severe depression)," psychologists Avshalom Caspi and Terrie Moffitt assert.[21]

Contrary to *DSM* assumptions, instead of a large number of distinct entities, mental disorders seemed more related to a small number of general vulnerabilities such as "psychoses," "internalized neuroses," and "externalized neuroses." These common factors make people prone to develop a variety of different conditions rather than separable disorders. This startling finding was in certain ways closer to DSM-I and II conceptions of mental disorder, etiology notwithstanding, than the numerous distinct categories that dominated subsequent DSMs.

An additional finding that emerged from neuroscientific research was the tremendous heterogeneity of the supposedly homogeneous DSM diagnoses. Major depressive disorder presented the clearest case. "There is every reason to believe," a review concluded, "that among the range of individuals currently subsumed under the diagnosis of Major Depression are those with distinct disease states mediated by very different underlying pathophysiological mechanisms."[22] Even the supposedly homogeneous condition that Kraepelin identified as dementia praecox turned out to be far more heterogeneous than the German pioneer believed. Although, as Kraepelin insisted, many people with schizophrenia show a deteriorating course as they age, large proportions stabilize, improve, or even recover as they grow older.[23]

The lesson from the large body of evidence that had accumulated since 1980 seemed clear: the DSM's categorical and theory-neutral structure hindered diagnostic progress. In contrast to the specific, discrete, and seemingly homogeneous entities found in the DSM, research was uncovering that mental disorders were broad, overlapping, and heterogeneous. The result was that the central goal that the Washington University pioneers, following Kraepelin, proposed for psychiatric classification—to develop homogeneous entities with distinct descriptions, explanations, and outcomes—remained almost entirely unrealized.[24] Psychiatry would have to change the essential character of the manual to take advantage of the breakthroughs that neuroscientific techniques promised. Like the DSM-III, the DSM-5 would revolutionize the diagnostic process and allow the field to fully partake in the latest scientific advances.

Identifying Problems with the DSM

The DSM-5 revision process began in 1999 when the APA and the NIMH jointly sponsored a conference devoted to setting research priorities for a manual that would replace the DSM-IV. Dissatisfaction with the extant diagnostic paradigm was apparent from the outset. The NIMH, which had been one of the most enthusiastic participants in developing the previous three manuals, had started to question the value of the DSM system itself. By the turn of the century, the agency had made a 180 degree turn away from its socially oriented postwar origins to embrace an organic model of mental illness. It was especially worried that the *DSM* did not incorporate biological and genetic findings that were emerging from the nascent field of neuroscience. Instead, the NIMH aspired to develop a system that would "introduce neuroscience into diagnosis."[25] Later, its then-director, Steven Hyman, noted, "I was spending taxpayer money on grants that were being forced into categories that might or might not conform to nature" and called the *DSM* "an absolute scientific nightmare."[26]

Even at this early stage, it was apparent that the goals of the new classification would significantly differ from those of prior revisions. Initially, the leaders of the *DSM-5* process believed that they could go beyond the descriptive approach of previous manuals to incorporate etiological findings from neurogenetic research. They optimistically predicted, "As new findings from neuroscience, imaging, genetics and studies of clinical course and treatment response emerge, the definitions and boundaries of disorders will change."[27] The manual's categorical diagnoses, however, presented barriers to achieving these goals. The initiators of the revision ridiculed earlier *DSM*s as "anachronistic pure types of previous scientific eras." Instead, the *DSM-5* would overturn the diagnostic history that ran from the 1972 Feighner criteria through the *DSM-IV*: "In retrospect, it is interesting that there was such a strict separation of mood, anxiety, psychotic, somatic, substance use, and personality disorder symptoms for the original Feighner diagnoses."[28] In a perhaps unconscious echo of the critique that Spitzer and others made of the *DSM-I* and *II*, they promised that the *DSM-5* would replace the existing manual with "diagnostic criteria that are intended to be scientific hypotheses, rather than inerrant Biblical scripture."[29]

The first phase of the revision involved numerous conferences, 13 in all. Biological psychiatrists dominated the initial stage of developing the manual. According to Michael First, a key participant in the *DSM-IV* revision, "[The APA leadership] had this idea that we had the wrong people doing the DSM, and if we had the neuroscientists and geneticists around, they'd tell us all we need to do to make a paradigm shift."[30] The early meetings identified a number of issues for the *DSM-5* planning process to address.[31] The first was the current manual's failure to incorporate the neuroscientific and genetic results mentioned above. In addition, the new classification must somehow take into account the findings regarding the current system's over-specificity, diagnostic overlap, heterogeneity, and overuse of not otherwise specified (NOS) diagnoses when impairing symptoms were present but did not fully meet any particular criteria set. Soon, however, interpersonal squabbling would come to overshadow these academic considerations.

In 2006 the APA selected David Kupfer, a leading academic authority on mood disorders, to chair the DSM-5 Task Force. Kupfer teamed with Vice Chair Darrel Regier, the APA's research director and a notable epidemiologist, to oversee the revision. In outward form, this process resembled the construction of the earlier manuals. Kupfer and Regier assembled a task force of about 30 members to oversee the procedure, 13 work groups to revise the major *DSM-IV* categories, and well over 100 consultants to the work groups. Yet, in practice, the development of the *DSM-5* turned out to be thoroughly distinct from previous revisions.

An Oedipal Crisis

The APA and the DSM-5 Task Force made a number of missteps as the revision proceeded. One was the tactical error of excluding the central figures who developed the previous manuals. "To encourage thinking beyond the current DSM-IV framework," the *DSM-5* leadership rejected not just the intellectual framework of previous manuals but also—echoing Spitzer's marginalization of psychoanalysts in the *DSM-III* deliberations—ostracized both Spitzer and his successor, Allen Frances, from the *DSM-5* process.[32]

The spark for the dispute was the revelation in 2008 that all work group members had signed nondisclosure agreements that forbade them from publishing or divulging the content of their discussions.[33] This was of particular concern to many critics because most members of these groups had financial ties to the pharmaceutical industry.[34] They feared that drug companies would have substantial but hidden influence over the revision process. Spitzer's request for the minutes of a work group meeting was denied because of these confidentiality agreements. Spitzer's outrage at the secrecy of the *DSM-5* revision led him to become one of the most fervent opponents of the entire process. He mocked Kupfer and Regier's claim that the *DSM-5* was "the most inclusive and transparent developmental process in the 60-year history of DSM," citing the lack of information about work group discussions, the confidentiality agreements that all members were forced to sign, the opacity of the appointment process for work group members, and the "astounding claim" that the new manual would be more etiological than the *DSM-IV*.[35]

Allen Frances, who had overseen the *DSM-IV* revision, was an even more vociferous opponent than Spitzer of the *DSM-5* process. Frances's major concern was the vast over-medicalization of ordinary emotions that he claimed the *DSM-5* would produce. He became the manual's most public and fervent critic, publishing a regular blog on the *Psychology Today* website, writing op-eds in the *New York Times* and other prominent newspapers, and making numerous television appearances.

While the organization's leadership could portray many of the manual's oppo-nents as "the familiar antipsychiatry critics," it could hardly do the same with Spitzer and Frances.[36] Nevertheless, then-APA president Alan Schatzberg and other APA leaders including Kupfer and Regier harshly counterattacked. In a clas-sic example of projection, they claimed the DSM-5 process was driven by purely "scientific" concerns while their opponents' motivations were mercenary. Schatz-berg (who had $4.8 million of stock options in a drug development company) and the others accused Frances (who was receiving about $10,000 in yearly royalties from the DSM-IV) of "failing to disclose his continued financial interests in sev-eral publications based on DSM-IV." They added, "Both Dr. Frances and Dr. Spitzer have more than a personal 'pride of authorship' interest in preserving the DSM-IV and its related case book and study products. Both continue to receive royalties on DSM-IV associated products."[37] The APA went to war with the psychiatrists who were most identified with the previous DSMs.

The organization went further than banishing Spitzer and Frances. "All the people at the top of the previous DSMs were completely excluded," Michael First observed.[38] The media reveled in the intensely personal disputes the DSM-5 pro-cess was producing. "What began as a group of top scientists reviewing the re-search literature has degenerated into a dispute that puts the Hatfield-McCoy feud to shame," a Psychiatric Times blog noted.[39]

The Field Trials

Spitzer was also involved in a dispute over the findings from the field trials that the task force conducted to establish the reliability of the proposed diagnoses. They were outwardly similar to those of previous revisions, which went virtually unno-ticed while they were conducted.[40] Unlike prior efforts, however, the tests of the projected DSM-5 diagnoses generated considerable controversy. The NIMH had funded the earlier research. The agency, however, refused to fund the DSM-5 trials, stating that it was no longer willing to use public money to support a manual that a private professional organization owned. This forced the APA to fund the field trials itself. The resulting process suffered from poor participation rates, contin-ual delays, and the abandonment of their most ambitious goals. By the time the trials began, little time was left until the draft manual had to be completed.[41]

The results of the field trials were surprisingly disappointing. As in the earlier trials, the DSM-5 procedure used the kappa statistic, a measure of the extent to which different raters agree on what diagnosis a patient should receive, as its mea-sure of reliability. Kappa values range from 1, where two raters completely agree on some diagnosis, to 0 where the extent of their agreement does not exceed random chance.[42] Previous studies had considered kappas of greater than 0.8 as excellent, 0.6–0.8 as good, 0.4–0.6 as fair, and those lower than 0.4 as poor or "perilously close to no agreement."[43] The kappas for the DSM-5 trials were startlingly lower than they had been. While a few diagnoses, including posttraumatic stress disorder, attention deficit/hyperactivity disorder, and autistic spectrum disorders, achieved kappas that would have been considered good in prior trials, most were in the fair range, and some, including the workhorse diagnoses of MDD and generalized anxiety disorder, were poor. The proposed diagnosis of mixed anxiety depression disorder attained an astonishing kappa of 0, the equivalent of monkeys throwing darts at a diagnostic target. Moreover, most of the kappas were considerably lower than those for equivalent diagnoses in the DSM-III and DSM-IV field trials. For example, kappas for schizophrenia declined from 0.81 in the DSM-III trials to 0.46 in the DSM-5 trials; in the same period the statistic for major affective disorder went from 0.80 to 0.25.[44] Finally, the trials indicated significant divergences across different research sites. The most heralded achievement of the DSM-III revolution, increased diagnostic reliability, seemed to have vanished.[45]

In essence, the task force tried to resolve this crisis by moving the goal posts. The head of the field trials, statistician Helena Kraemer, mocked the DSM-III and IV trials: "There was a bias in those studies" leading their findings to be "badly inflated and that causes a problem now."[46] The DSM-5 trials redefined kappas above 0.8 as "miraculous," those between 0.6 and 0.8 as "cause for celebration," ones between 0.4 and 0.6 as "realistic," and findings between 0.2 and 0.4 as "acceptable."[47] They thus claimed that the "substantial majority" of assessments had "good or excellent" results.[48] For example, a kappa of 0.54 for borderline personality indicated "a major step forward." The 0.25 kappa for the new diagnosis of disruptive mood dysregulation disorder was "more modest."[49] Overall, the reliabilities, which previous standards would have considered to be poor or fair, were even more distressing because most were obtained in specialty academic settings rather than in regular psychiatric outpatient or primary medical care practices.

Spitzer derided the results of the field trials as "a step backward" from those in the DSM-III.[50] In turn, Kupfer and others involved in them ridiculed the DSM-III trials that Spitzer led because of their selection bias, lack of clinician blindness to patient diagnosis, and small sample sizes.[51] The exchange did not inspire confidence in either set of trials. The justification for the DSM structure, which invoked high reliability as its key basis of legitimacy, was disintegrating.

A Chaotic Process

The construction of the *DSM-5*, which involved a chair, a task force, different work groups, and a variety of consultants, echoed previous revisions. However, the organization of the *DSM-5* differed from prior manuals in a major way. Spitzer had closely supervised every aspect of writing the *DSM-III* and *DSM-III-R*; no change occurred in either without his assent. His successor for the *DSM-IV*, Allen Frances, was not as hands on as Spitzer and gave more authority to the particular work groups involved. All of these groups, however, followed common standards that embodied a conservative strategy assuming existing diagnoses would be retained unless overwhelming empirical evidence dictated some change.[52]

In contrast, Kupfer and Regier encouraged each work group to develop its own proposals that need not be based on existing diagnoses. "Kupfer and Regier gave nearly complete autonomy to the workgroups, initially telling them that the sky was the limit and they should not treat the *DSM-IV* syndromes as inerrant scripture," a history of the process observes.[53] Different groups would develop their own strategies for modifying the class of disorders they considered. The result was that those working on the *DSM-5* revision were both far less committed to retaining the content of the manual they were altering and their process was far more decentralized.

The decentralization of the *DSM-5* revision meant that the 13 work groups developed highly distinct proposals. The diversity that emerged threatened to undermine the *DSM-III*'s hard-won revolution that standardized the diagnostic process. Many members reported frustration with the lack of direction from task force leadership. One of the participants in sociologist Owen Whooley's study reported, "I get aggravated with Kupfer and Regier [chair and cochair of the task force] sometimes, where I want to say, 'For God sakes, you have to tell us how many dimensions we can have.' I mean these are things where you really need somebody to make the decision about what the parameters are so that you can work. These guys are just way too open and flexible for us."[54] In addition, while the major controversies in the making of the *DSM-III* involved analytic outsiders who opposed empiricist insiders, the *DSM-5* process was marked by internal divisions within committees. Many workgroups featured in-house arguments among their members; several resigned in disgust over the deliberations.[55] In the end, some of the workgroup plans succeeded, some failed, and others were tabled.

In 2009 the extreme decentralization of the *DSM-5* process, the dissension that developed in many of its groups, and the very public controversies that emerged led the APA leadership to appoint an oversight committee to supervise all the work groups. The board of trustees was especially concerned with the disorganization of the procedure and the lack of progress the task force was making. "There is a serious problem with the DSM, and we've got to fix it," the committee chair reported.[56] It backed away from the initial goal to implement a paradigm change and placed restraints on the degree of change in the new manual. A year later it established a Scientific Review Committee (SRC) that was independent of the DSM revision structure to oversee all the proposed changes to the manual and make recommendations directly to the APA president and board of trustees. In a reversal of the early charge to the work groups that "the sky was the limit" for changes, the SRC's mandate was to set "a high bar for major changes."[57] The SRC's rejection of ambitious proposals to create a new diagnosis of "psychosis risk syndrome" and, especially, to make continuous rather than categorical measures of the personality disorders illustrate how the DSM-5 process eventually resulted in retaining the diagnostic status quo.

The Psychosis Risk Syndrome Proposal Fails

One contentious controversy during the run-up to the DSM-5 was whether the manual should contain a diagnosis of psychosis risk syndrome (PRS). Psychiatry had long sought to develop some way of detecting and treating incipient forms of psychoses before they developed into full-blown cases. For example, in 1895 neurologist Silas Weir Mitchell critiqued as "superstitions" the assertions of asylum psychiatrists: "You hear the regret in every report that patients are not sent soon enough, as if you had ways of curing which we have not."[58] In the 1920s and 1930s the mental hygiene movement promoted the idea that identifying and treating at-risk children and adolescents could prevent them from ever becoming psychotic. More recently, an influential crusade had arisen to screen every youth for early signs of mental illness with scales having as few as two items.[59] Yet, despite this history of attempts at prevention, no previous manual contained any diagnosis that identified people who didn't have a mental disorder but might develop one at some future point.[60] The DSM-5 Psychotic Disorders work group (PDWG) tried to change this situation. Had the group's proposal succeeded, it would have been a revolutionary development in psychiatric diagnosis.

Advocates for incorporating PRS into the DSM-5 presented several rationales. For one thing, the psychoses, schizophrenia in particular, were the most devastating form of mental disorder. Most first emerge in late adolescence or early adulthood, so their identification at an early phase might prevent many decades of later suffering for those afflicted and their families. Studies of young people who had sought some sort of psychiatric treatment indicated that, on average, about 18 percent developed a serious psychiatric disorder within six months and about 36 percent after three years. Many adolescents, however, were not getting an appropriate diagnosis that indicated their condition might deteriorate.[61]

In addition, proponents claimed that mild forms of mental disorder often preceded the development of more serious ones.[62] Longitudinal studies of the general, untreated population indicated that people who had severe conditions at a later point had earlier shown milder signs of disorder. Therefore, the presence of minor symptoms could be predictors of severe conditions. Finally, PRS advocates noted that many interventions in general medicine tried to prevent diseases that were not yet apparent from ever arising. For example, millions of people with elevated levels of cholesterol or blood pressure take medication aimed at averting future heart attacks. There was no reason psychiatry shouldn't strive to do the same.

Based on these justifications, the PDWG developed criteria for a new diagnosis to identify those who would likely become psychotic in some future period. PRS required at least one "attenuated" symptom of delusions, hallucinations, or disorganized speech at least once a week for the previous month. Such symptoms did not meet the full criteria for any psychotic diagnosis but were significant enough to lead to distress among patients or those around them or to lower functioning.[63] The chair of the work group, William Carpenter, argued, "Without the [diagnosis], a framework for early detection and intervention is lost, and the opportunity to develop evidence-based treatment will be minimized." A PRS diagnosis would also "facilitate evolution in the understanding of disease trajectories and refinement of early treatment."[64]

The psychosis risk proposal generated ferocious opposition, including some from the work group's own members. Critics especially worried that this new type of condition would create an attractive target for drug companies to exploit. PRS appealed to the pharmaceutical industry because it usually arose among adolescents and seemed to require lifetime regimens of antipsychotic medication. Allen Frances was the leading adversary. He was unsparing in his critique, noting that at least two-thirds of people in treatment who received the diagnosis would *not* go on to develop a psychosis. Worse, in untreated groups the number of false positives would be around nine times higher than correctly labeled persons. Moreover, despite the good intentions of its advocates, no treatment could prevent persons who were identified as at risk for developing a psychosis because antipsychotic medications showed no benefits before overt psychotic symptoms arose.[65] Finally, everyone who received a PRS diagnosis—both the majority who never actually developed a psychosis as well as the minority who did—would be stigmatized. Frances concluded, "If 'psychosis risk' made it into DSM-5, innocent kids might become obese and die early receiving unnecessary medication for a fake diagnosis."[66] Advocates themselves recognized that the dangers of a PRS diagnosis included potential stigmatization, overmedication, and false positives but argued that the potential benefits from inclusion outweighed these costs.[67]

The harsh criticisms, dissension among work group members, and lack of a solid evidence base led the SRC to reject the PRS proposal. Instead, the PDWG changed the name of the proposed diagnosis to attenuated psychosis syndrome (APS). The name and accompanying criteria eliminated any references to risk or prevention and instead were purely based on *current* symptoms that did not rise to the threshold of a full psychosis.[68] Second, the *DSM-5* placed APS in the section on "conditions for further study" so that it was not officially recognized for use in clinical settings. This diagnosis will undoubtedly generate much research for, as Frances observed, if efforts to identify and treat schizophrenia at an early stage ever succeed, they "would be the highest achievement in the history of psychiatry."[69] Yet, for the time being, this ambitious attempt at radical diagnostic change had failed.

THE DIMENSIONAL REVOLUTION COLLAPSES

By the time the *DSM-5* work groups began meeting in 2008, it had become apparent that they would not be able to implement a biologically informed diagnostic system.[70] Studies indicated far weaker genetic evidence than researchers had expected. In addition, they had not uncovered biomarkers for any mental disorder, and the prospects for finding them were not promising. Therefore, a different issue came to the fore: perhaps mental disorders weren't categories at all, but, instead, "disorders might merge into one another with no natural boundary in between."[71] The DSM-5 Task Force discarded its initial goal of infusing biological information into the manual and replaced it with a plan to implement a dimensional system to supplement and eventually replace the *DSM*'s categorical classifications.

Dimensional measurement of mental traits was a common practice in the field of psychology but a rarity within psychiatry. For decades, psychologists had developed sophisticated numerical scales that assumed mental traits varied continuously along dimensions such as intensity, severity, and duration. Psychologists didn't think in terms of cutoff points between disorders and non-disorders but instead conceived of points on quantitative measures. People received a precise score on each trait that placed them in relationship to the scores of others on the same trait. The best-known measures were the "Big Five" personality dimensions of openness to experience, conscientiousness, extraversion, agreeableness, and neuroticism.[72] Everyone was more or less extroverted or introverted, agreeable or antagonistic, conscientious or impulsive, and so on. Measures of personality provided seemingly exact and quantified results that emerged from rigorous empirical studies.

Numerous researchers argued that instead of a diagnostic dichotomy between those who had some diagnosis and those who did not, many *DSM* categories actually reflected positions on dimensions with no inherent cut points.[73] Rather than classifying a disorder as present or absent, a new system would view mental disorders as matters of degree that ranged from not at all severe to very severe, with mild and moderate conditions falling in between. Each place on this range would be given a numerical value on a quantitative scale. Proponents of dimensions anticipated that the new *DSM* would increase validity through using them in addition to—or even instead of—categories. Moreover, identifying less-severe symptoms provided the opportunity to intervene before they progressed to a more serious state. Dimensional measurement would be "the single most important precondition for moving forward to improve the clinical and scientific utility of DSM-V."[74]

The class of personality disorders seemed ideally suited for implementing the task force's dimensionality aspirations.[75] An extensive body of psychological research considered all personality traits as continuous: those who were considered disordered were on the extreme ends of dimensions that contained everyone. Presumably, the same methods could be used for the other major classes of mental disorder. In 2004 this category had been the subject of the first research conference held on the *DSM-5* revision. Converting the personality disorders into continua could be "foundational to the conceptualization of psychopathology in general."[76]

The composition of the Personality Disorders work group (PDWG) indicated the task force's determination to create radical change: only one member was a holdover from the *DSM-IV* revision. Another sign of its desire to thoroughly revamp the category was that a majority of the work group's members were not psychiatrists but held PhDs in psychology. The task force didn't seem to be concerned that personality psychologists don't think in terms of diagnoses—or even of actual people—at all. Their worldview could hardly be more distinct from encounters between therapists and patients. Personality psychologists do not engage with individuals in clinical practice but examine numbers on computer screens. They use procedures such as rotating factors and developing covariant attribute clusters that produce the best statistical fit for their data. In his classic critique of the field, psychologist Walter Mischel summarized, "It is as if we live in two independent worlds: the abstractions and artificial situations of the laboratory and the realities of life."[77] Unsurprisingly, the proposed marriage between psychology and the *DSM-5* was short and conflictual, and it ended badly.

The PDWG viewed the personality disorders as a test case for showing the advantages of transforming categorical into continuous conditions.[78] It embraced its mission as the pioneer for a dimensional system that could eventually serve as a model for the entire *DSM*. Work group members were convinced that what previous *DSM*s considered disorders were actually extremes on dimensions that applied to everyone. They also believed that the same complex statistical models that yielded personality types in the general population would work with clinical groups.

By 2009, the PDWG had developed an ambitious plan to thoroughly revamp the personality disorders and eliminate diagnostic thresholds. Its central recommendation was that the *DSM-5* should implement a multilevel assessment with subdivisions of individual and interpersonal personality functioning. Its proposal featured a new four-part assessment that used a four-point scale for each of five broad personality-trait domains of negative affectivity, detachment, antagonism, disinhibition, and psychoticism. The scale would rate each respondent's degree of similarity to these five personality dimensions as well as the presence of traits organized into six domains and 37 facets.[79] This multidimensional procedure would produce a unique score for every assessed individual. Without irony, the *DSM-5* website proclaimed, "The proposed system is designed for flexible use to maximize clinical utility."[80]

Unsurprisingly, the work group's proposal for the personality disorders generated immediate and fierce opposition. Ironically, the first principle of the *DSM-5* revision was that "the DSM is above all a manual to be used by clinicians, and changes made for DSM-V must be implementable in routine specialty practices."[81] Yet clinicians responded with horror to the amazingly complicated proposal. They were especially concerned that the "voice of clinicians appears to be particularly lacking."[82] Observing how the research the committee relied on involved quantitative studies of untreated populations by academic psychologists, they questioned its usefulness for actual practice. "For all their scholastic erudition, the work group have created a monster—a bloated, pedantic, cumbersome diagnostic instrument that will never be used by anyone working in the hurly-burly of clinical practice," psychiatrist James Phillips noted.[83] Many were also angered that the PDWG recommended keeping diagnoses for just 6 of the 10 existing personality disorders (antisocial, avoidant, borderline, narcissistic, obsessive-compulsive, and schizo-typal but not paranoid, schizoid, histrionic, or dependent).[84] Other critics empha-sized the huge discontinuity between the *DSM-5* proposals and the extant *DSM-IV* criteria.

In May 2010, 29 prominent clinicians and researchers addressed a letter to the PDWG, DSM-5 Task Force, and APA trustees objecting to the revisions.[85] Other critiques appeared in psychiatry journals and on the *DSM-5* website. The revision procedure also featured powerful internal divisions within the work group. Al-though all committee members were committed to dimensional measurements, they disagreed intensely over what particular model best captured the nature of the personality disorders. In the face of massive clinical opposition, as well as sharp disagreements among work group members, the Scientific Review Commit-tee rejected the PDWG proposal. After a complex series of reviews, the APA Summit Committee also overruled the plan. While the *DSM-5* abolished Axis II and thus integrated the personality disorders with other mental disorders, the manual did not incorporate any of the work group's other aspirations.

The amazing result was that the *DSM-5* simply retained, word for word, the personality disorders section from the *DSM-IV*. In a face-saving move, the work group's dimensional proposals were placed in a new section on emerging measures and models called "Alternative DSM-5 Model for Personality Disorders." It pre-sented the multidimensional model and scoring system for six personality condi-tions (antisocial, avoidant, borderline, narcissistic, obsessive-compulsive, schizo-typal).[86] The criteria, however, were not codable diagnoses but were instead analogous to the *DSM-IV* conditions for further study. As the major players in the personality disorders work group explained about the debacle, "It is a story of shift-ing expectations, conflicting goals, and fractured alliances."[87] Theodore Millon, a member of the *DSM-IV* Personality Disorder work group went further, calling the whole process "nothing less than a sorry mess."[88] The workaday concerns of cli-nicians had thwarted the grand aspirations of researchers.

Dimensionalization was also a step too far for the APA membership. The task force's plans to introduce dimensional scales to screen all incoming patients and to measure the severity of all the manual's conditions failed. Clinicians, who were the most sizable constituency of the organization, were not familiar with the use of dimensions, found them burdensome and hard to understand, and worried about their implications for reimbursement. The relationship between some point on a continuum and a diagnosis seemed opaque and arbitrary. Moreover, clini-cians were hardly involved in the actual revision process at all. Whooley's research indicates that the primary affiliations of all but 2 of the 31 task force members were with a university or a research institute. Each of the 13 work group chairs identified as a psychiatric researcher, as did over 95 percent of the 160 individuals involved in the revision.[89]

The invisibility of clinicians in the DSM-5 revision had major repercussions when the proposed manual came up for the APA's approval. At the annual meeting in May 2012, the APA Assembly, a body mostly composed of clinicians selected by local psychiatric societies, voted unanimously to relegate all the dimensional scales to the appendix of the manual. It not only saw them as a threat to clinical practice but also believed that dimensional assessment was far from ready for implementation. The APA Board of Trustees affirmed this decision. With the proposals for dimensional measurement safely tucked away in an appendix, the APA approved the DSM-5.[90]

The task force suffered an even worse blow a week before the new manual was to become operational in May 2013. The NIMH, which had been the APA's steadfast ally in conceiving and implementing all previous manuals, disavowed the entire DSM system. NIMH director Thomas Insel dismissed the new DSM-5, noting that "the final product involves mostly modest alterations of the previous edition."[91] Worse, Insel derided the DSM system: "In the rest of medicine, this would be equivalent to creating diagnostic systems based on the nature of chest pain or the quality of fever."[92] Insel went on to say, "Patients with mental disorders deserve better" and announced that the NIMH was developing a completely new dimensional system, the research domain criteria (RDoC), which would be based on the biology of brain circuits. Because "it is critical to realize that we cannot succeed if we use DSM categories as the 'gold standard,'" the NIMH would reorient its research from the presumably antiquated DSM categories to the new RDoC system.[93] The pedestal the DSM mounted in 1980 had seriously cracked and was in danger of completely shattering.

Regressions in Psychiatric Diagnoses

The rejection of its major innovations crushed the high aspirations of the DSM-5 Task Force. The paradigm shift it envisioned did not materialize. Instead, the DSM-5 looks much like its predecessors. It failed to make progress on the problems the task force had identified: psychiatry's inability to establish etiologically based conditions, develop dimensional measurements, reduce the overlap among diagnoses and the heterogeneity of many of the DSM entities, or create a diagnosis that could predict the emergence of severe conditions before they arose. Moreover, some of the changes that the manual implemented did not improve but actually reduced the validity of a few of its major diagnoses. Major depressive disorder and substance use disorder (SUD) illustrate this dynamic.

Increasing Invalidity: Major Depressive Disorder

The *DSM-III* revolution in 1980 had made MDD psychiatry's most prominent condition, accounting for far more diagnoses than any other *DSM* entity. This resulted from its heterogeneous criteria, which did not distinguish people who became moderately depressed because of inherent biological or developmental vulnerabilities to precipitating life circumstances from those with debilitating melancholic symptoms who were unresponsive to life events. Moreover, the two-week duration requirement also captured a third category of people who developed brief periods of normal sadness following losses of relationships, jobs, and health, as well as many other stressors.[94] Its three-pronged character made MDD so diverse that researchers could not make progress in isolating the etiology, prognosis, or effective treatments for the most common mental disorder.

The MDD diagnosis faced two different kinds of attacks during the *DSM-5* process. The first came from biologically oriented researchers, who argued that melancholic depression was a discrete entity that deserved its own diagnosis in the *DSM-5*. A group of 17 distinguished psychiatrists and others noted the distinctive nature of melancholia in an editorial in the *American Journal of Psychiatry*.[95] It featured unresponsiveness to external changes; profound hopelessness; slowness of thought, movement, and speech; and vegetative patterns of sleep, appetite, and libido. Melancholic depression was also likely to be a lifelong condition with many recurrent episodes. Unlike more common forms of MDD, it responded well to tricyclic antidepressants and electroconvulsive therapy but not to placebos or SSRIs. Finally, the dexamethasone suppression test, which measures levels of cortisol hormones, could sometimes detect it.

The signers of the editorial were a virtual who's who of depression researchers, including Spitzer and Donald Klein, prominent biological psychiatrists, and even analytic psychiatrists and a historian. Their aggregate accomplishments as biological researchers dwarfed those of the members of the MDWG. Moreover, their proposal seemingly embodied exactly the kind of organically grounded diagnosis that the *DSM* strove, but failed, to incorporate since 1980. Finally, since the time of the Hippocratic physicians, medicine had recognized the kind of melancholic syndrome the diagnosis isolated. Even the much-derided *DSM-I* and *II* separated psychotic from neurotic forms of depression. The evidence base, prominence of supporters, and historical grounding seemed to guarantee the inclusion of melancholia in the *DSM-5*. As prominent psychotherapist Gary Greenberg observed, a melancholic diagnosis would have shown "that, at least in this one case, psychiatrists were real doctors treating real diseases that could be discerned with real tests and treated with real cures."[96]

The MDWG, however, was unimpressed and disregarded the proposal for a melancholia diagnosis, which does not appear in the *DSM-5*. Greenberg speculates about its motive: "A test for melancholia would make the lack of biological measures elsewhere in the DSM that much more glaring. It was a success that would only highlight the APA's failures."[97] Another possible reason for the work group's decision was that separating melancholy from MDD would have abandoned the "major" part of the diagnosis. What remained within MDD could have become more vulnerable to accusations that its criteria medicalized normal emotions of sadness. The work group's rejection of a proposal from many of the most prominent figures in depression research protected psychiatry's interests in making its central condition look like a truly serious and legitimate disorder.

A very different line of attack on MDD was pursued by philosopher Jerome Wakefield and Allen Frances, among others. Their critiques built on the only notable exception to the acontextual MDD criteria. The *DSM-III-R* and *IV* definitions of mental disorder recognized that "an expectable and culturally sanctioned response to a particular event, for example, the death of a loved one" was not a mental disorder.[98] Accordingly, MDD excluded conditions that arose after the death of a beloved person that otherwise met diagnostic criteria. People with what the manual called "uncomplicated" depression would not receive a diagnosis as long as they did not have an especially severe symptom (psychotic symptoms, suicidal ideation, psychomotor slowness, preoccupation with worthlessness, or marked functional impairment) or extended duration (at least two months instead of two weeks). This group contrasted with "complicated" conditions among the bereaved that did have one or both of these qualities. Yet the bereavement exclusion (BE) was the sole exception to the MDD criteria, not an "example" of a broader category of loss that the general definition of mental illness portrayed.

Wakefield, in particular, argued that grief was not unique but should serve as a model for all kinds of losses, for example, close relationships, jobs, finances, health. He and others conducted a series of empirical studies that showed how all kinds of uncomplicated losses had similar symptoms, duration, treatment history, and degree of impairment as uncomplicated bereavement. In addition, the psychic damage from a range of losses that met MDD criteria but were not complicated naturally dissipated and did not return after the loss was resolved. Indeed, over time, the rate of recurrence for people who suffered from loss-related depressions was more similar to people who were never depressed than to those with complicated conditions.[99]

Based on these empirical findings, critics contended that the crucial distinction was not between bereavement and other losses but between uncomplicated conditions following loss and complicated conditions that featured especially severe symptoms or prolonged duration. Proponents of expanding the BE made two central points: first, there was no substantial difference between the impact of bereavement and other losses on depression and, second, that many uncomplicated loss-related depressions were not depressive disorders at all but were natural responses to loss.[100] Therefore, they argued that the exclusion should also apply to losses other than bereavement.

The attempt to extend the bereavement exclusion to a greater range of losses met an even worse fate than the proposal for a new diagnosis of melancholy did. The work group considered only the claim that there was no reason to single out bereavement as a special kind of stressor. Therefore, it used the same argument that Wakefield and others used as a rationale—there was nothing unique about bereavement—to *eliminate* the exclusion from the text of the diagnosis. "The exclusion of symptoms judged better accounted for by Bereavement is removed because evidence does not support separation of loss of loved one from other stressors," the MDWG wrote.[101] It ignored the second, and more basic, argument that would broaden the BE to other kinds of losses.

In place of the discarded BE, the work group added an explanatory note and a footnote that were not part of the MDD diagnostic criteria. The note stated that the diagnosis "requires the exercise of clinical judgment based on the individual's history and the cultural norms for the expression of distress in the context of loss."[102] The admonition to use "clinical judgment" as the standard for when loss-related symptoms represented a disorder undercut the standardization rationale that was a core principle behind the *DSM-III* revolution. The elimination of the BE also explicitly contradicted the *DSM-5*'s *own* definition of mental disorder, which uses bereavement as an example of a non-disordered condition: "An expectable or culturally approved response to a common stressor or loss, such as the death of a loved one, is not a mental disorder." Moreover, excising the BE made the already heterogeneous MDD criteria even more diffuse and prone to false positive errors.

Unsurprisingly, the elimination of the BE generated perhaps more negative publicity than any other change in the *DSM-5*.[103] No one needed professional knowledge to understand that non-severe and brief periods of sadness after the death of a loved one were not mental disorders. This decision also contradicted the task force's own principle that "a broad consensus of expert clinical opinion would generally be expected for all proposed changes or additions to DSM-V."[104] The work group's rhetoric that invoked evidence-based decisions rang hollow in the wake of a change that flew in the face of research findings and made an already indefensible diagnosis worse. MDD, the core diagnosis of psychiatry since 1980, became an even more incoherent combination of severe melancholia, depressive disorder, and normal sadness.

What led the work group to abandon the BE? Its manifest reason was the concern that bereaved people would not be able to receive helpful treatment.[105] Yet this rationale ignored the fact that the existing criteria *already* took into account that cases with either just a single severe symptom or a two-month duration would receive an MDD diagnosis. Instead, the best explanation for removing bereavement from the diagnostic criteria might be the genuine danger that expanding the exclusion would pose to the entire MDD edifice. If the failed proposal to establish a distinct diagnosis of melancholia would have chipped away at MDD, the expansion of the BE might have dismantled it. A large proportion of outpatient mental health visits stems from stress-related problems that result in MDD diagnoses. Extending the BE to other losses could have threatened psychiatry's legitimate jurisdiction over normal-range losses of all sorts. Eliminating the exclusion not just maintained but also increased psychiatry's authority to deal with the psychic consequences of psychosocial stressors. Ultimately, the professional benefits of maintaining a broad MDD diagnosis trumped the rhetoric of implementing an evidence-based diagnostic system.

Increasing Invalidity: Substance Use Disorders

For centuries, compulsive use of alcohol and drugs was viewed as a moral transgression, a sin, a legal problem, or all three. Over the past 50 years, however, the idea grew that addiction is a biological disease that warrants medical treatment instead of punishment. An addict is someone who habitually consumes some substance despite serious negative consequences. Google nGram analysis indicates that between 1960 and 2005 the use of the terms "addiction" and "addictive" expanded enormously.[106]

The *DSM-I* and *II* briefly defined several types of organic and psychogenic addictions to alcohol and some types of drugs.[107] Because none of these diagnoses involved analytic assumptions, substance use escaped the major impacts of the

DSM-III revolution. The central change in the *DSM-III*, which persisted through the *III-R* and *IV*, was to split substance use into two separate categories of dependence and abuse. The *DSM-IV* criteria for dependence required the presence of three or more out of seven possible symptoms, which included the two physiological dependence symptoms of growing tolerance to the substance or maladaptive problems from withdrawal, and five behavioral symptoms, including giving up activities, continuing use despite harmful effects, trying and failing to stop, taking larger amounts than intended, and spending significant time obtaining and taking the substance.[108] Research indicated that the substance dependence diagnosis had good reliability and validity in capturing the essential qualities of addictive behavior.[109]

The *DSM-IV*, however, made the diagnosis of substance abuse that involved "impairment in social or occupational functioning due to substance use" much more problematic. Unlike the definition of dependence, which tried to define the intrinsically dysfunctional properties of substance use, abuse referred solely to the adverse social consequences of use.[110] It required just one symptom of abuse, such as substance-related legal problems or arguments with spouses about intoxication. Therefore, the *DSM-IV* contained one diagnosis of substance dependence, which captured the general nature of addiction, and another of substance abuse, which violated the *DSM* principle of avoiding diagnoses that solely rested on harm in the absence of an underlying dysfunction, on social deviance, or on conflicts between an individual and society. A resolution of this situation seemed to involve correcting or eliminating the invalid abuse category. The *DSM-5*, however, moved in a different direction.[111]

The primary change in the renamed class of substance-related and addictive disorders was to abandon the division between substance dependence and substance abuse that had persisted in every manual since the *DSM-III*. In its stead, the manual substituted the single category of substance use disorders, which combined criteria for the previous two diagnoses into a single entity.[112] The head of the *DSM-5* work group responsible for the new diagnosis attributed the change to an "overwhelming abundance of evidence in favor of combining the abuse and dependence criteria."[113] The *DSM-5* criteria, however, seem to be a step backward from previous formulations of substance use disorders. The new diagnosis contained 11 possible symptoms, 7 from the earlier dependence diagnosis, 3 from the earlier abuse diagnosis, and a new symptom of "craving."[114] The work group justified this change because statistical analyses indicated that symptoms of dependence and abuse were highly correlated.

One problem with the *DSM-5* criteria is that the new diagnosis is much easier to obtain than the one it replaced and so is subject to many false positive results. Because SUD contained more possible symptoms than did the *DSM-IV* diagnosis, it seemed reasonable that the *DSM-5* criteria would also increase the number of symptoms a diagnosis required.[115] Instead, they decreased the number of necessary symptoms for SUD to just two. One review predicted that the lowered criteria would produce a 62 percent increase in SUD diagnoses in the general population compared to the *DSM-IV* criteria.[116]

A second problem was that the new unitary diagnosis incorporated the invalidity problems of the old substance abuse diagnosis. A person whose substance use led to social or interpersonal problems and the inability to fulfill major role obligations met the revised criteria. For example, someone who argues with their spouse about smoking in bed could have a "tobacco use disorder."[117] Changing patterns of law enforcement, legislation, and social attitudes might all influence the prevalence of the new diagnosis: Localities that established checkpoints for drunken driving could at the same time be increasing levels of a mental disorder. A law that legalized medical or recreational use of marijuana would simultaneously decrease rates of mental disorders in that state. Someone could become mentally ill by moving from one state to another! Instead of having one valid and one invalid substance use condition, as in the *DSM-IV*, the *DSM-5* criteria had a single diagnosis that could often be invalid.

Third, the new diagnosis was far more heterogeneous than those in previous manuals. The SUD diagnosis hopelessly confounds dysfunction and social deviance.[118] It combines in a single entity someone who persistently uses large amounts of a substance, is unable to stop using, and has constant cravings for it with someone who oversleeps and misses an exam after a night of heavy drinking.[119] Moreover, the latter type of case outnumbers the former, resulting in misleadingly high estimates of SUD compared to the *DSM-IV* dependence category.

The *DSM-5* changes to substance use disorders have led the area to become, in the words of one of its leading practitioners, Griffith Edwards, "a field in disarray."[120] They blurred the distinction between mental disorder and social deviance and conflicts between individuals and society, created the potential for large numbers of false positives, and led to greater diagnostic heterogeneity. These changes do not just refute proclamations that diagnostic changes reflect growing knowledge but also show how diagnoses can regress from their portrayals in earlier manuals.

Responding to External Pressures: Autism Spectrum Disorder

The controversies over PRS, the personality disorders, MDD, and SUD rarely involved those who received these diagnoses. In contrast, the central players in debates over the proposed autism spectrum diagnosis were patients and their families. Psychiatric disorders among children and adolescents have several distinctive properties. First, patients themselves seldom initiate help seeking for their perceived problems. Instead, their parents or, more rarely, their teachers or other adults obtain treatment for them. Second, many parents participate in well-organized and politically active organizations that powerfully influence the diagnostic process. Finally, these groups positively value psychiatric diagnoses, forcefully advocate for them, and resist any changes that would limit the services they receive.

The attempt to dimensionalize the *DSM-5* diagnoses led to the new diagnosis of autism spectrum disorder (ASD). The Neurodevelopmental work group (NWG) believed that the two *DSM-IV* diagnoses of autism and Asperger syndrome (as well as Rett's disorder, childhood disintegrative disorders, and pervasive developmental disorder not otherwise specified) were not separate categories. Instead, they were aspects of a single ASD that varied along the dimensions of deficits in social interaction, impairments in communication, and repetitive behavior patterns. Each dimension had three levels of severity ranging from "requiring support," to "requiring substantial support," to "requiring very substantial support."[121] In essence, the former Asperger syndrome was a milder version of autism.

The NWG proposal generated as much opposition as any in the *DSM-5* process. Autism had been growing at a stunning rate: its prevalence had increased by 119 percent from 2000 (1 in 150) to 2010 (1 in 68).[122] Most observers did not believe that these numbers reflected a rise in the actual amount of autism but attributed this upsurge to such factors as greater awareness of the condition and parents taking advantage of better special education services in schools for students diagnosed with autism.[123] Parents did not want their children to lose these diagnoses and accompanying benefits. Yet researchers who applied the proposed *DSM-5* criteria to the data collected for the *DSM-IV* field trials found startling findings. Only about 60 percent of those who met *DSM-IV* standards would also qualify for a *DSM-5* diagnosis of ASD. Just a quarter of Asperger patients would keep their diagnosis under the new criteria.[124] The proposed changes were an unusual case where a *DSM-5* diagnoses would entail more stringent criteria.

The ASD proposal led to a public outcry after a front-page story in the *New York Times* described the impending reduction in diagnoses.[125] Regier reported receiving more than 10,000 emails in response to the article.[126] Well-organized and politically active groups of parental advocates and, in the case of Asperger, subjects themselves mobilized to protest the proposed changes. These groups emphasized how diagnoses led to resources including special education, mental health treatment, and disability payments. They were concerned that the new ASD proposal could limit eligibility for these valuable services. Many in the "Aspies" community worried that placing them on the same spectrum as those with autism would be stigmatizing. Yet they also wanted to make sure that those among them who were receiving treatment and monetary benefits would continue to get them.[127]

While critics of the *DSM* typically worried about false positives and epidemics of mental disorders resulting from the medicalization of normal emotions, family members and individuals who actually received autism and Asperger syndrome diagnoses were concerned that they would *not* qualify for one. The *DSM-5* made a unique response to the advocacy groups concerned with these conditions. It implemented the new ASD diagnosis but added a note stating, "Individuals with a well-established DSM-IV diagnosis of autistic disorder, Asperger's disorder, or pervasive developmental disorder not otherwise specified should be given the diagnosis of autism spectrum disorder."[128] Because these diagnoses had become such valuable commodities, a grandfather clause was created so that no one would lose them even if they no longer met criteria.

Whether the new ASD diagnosis represents progress from the former distinct categories is an open question at present. It does overcome the vast heterogeneity of the *DSM-IV* categories it replaced: one group of experts calculated that it reduced the 2,027(!) ways to qualify for the former conditions to just 11 possible combinations.[129] It also provides a test case for the value of a dimensional system. What the *DSM-5* deliberations over ASD unquestionably show is the extent to which external interest groups influence the putatively evidence-based diagnostic process. Many diagnoses have become so socially embedded that it is virtually impossible to eliminate them from the *DSM*.

Responding to External Pressures: Gender Dysphoria

Perhaps even more than the autistic spectrum proposal, the gender dysphoria diagnosis illustrates the extent to which *DSM* diagnoses have become invaluable resources. Strong opposition from gay advocacy groups forced the *DSM-II* to drop the homosexuality diagnosis. After that embarrassing controversy, the *DSM-III* and

III-R limited identifiable disorders of sexual orientation to people who were distressed about their conditions, before the *DSM-IV* abandoned the category alto-
gether. The issue took an interesting twist with the development of the *DSM-III*
category of gender identity disorders (GID). This diagnosis was applied when there
was "incongruence between anatomic sex and gender identity" and when one had
a persistent wish to "be rid of one's genitals and to live as a member of the other
sex." The *DSM-III-R* and *IV* elaborated the *DSM-III* diagnostic criteria but did not
fundamentally change them.[130]

The dynamics involving the GID diagnosis in the *DSM-5* process were virtually
the opposite of those concerning homosexuality. In contrast to their criticism of
this diagnosis in earlier *DSMs*, the major source of contention leading to the *DSM-5*
was advocates' insistence that the manual *maintain* a diagnosis related to cross-
gender identification. Ironically, the National Gay and Lesbian Task Force, which
had led the opposition to the homosexuality diagnosis in 1973, was the most vocal proponent of keeping the existing GID criteria.[131]

When the *DSM-5* work group on sexual and gender identity disorders appointed
two psychologists who questioned the GID diagnosis, transgender groups pro-
tested that some members of the work group believed that cross-gender identification was learned rather than innate.[132] Gary Greenberg notes "the irony of plead-
ing with psychiatrists *not* to take away a diagnosis that explicitly pathologized an
inborn condition."[133] Advocates were also worried about the practical need to obtain payment for a medical operation: transgender people who desired sex change
operations required a diagnosis before insurance companies would reimburse en-
docrinologists and surgeons who participated in gender transition surgery.

In response to advocates' concerns, the *DSM-5* kept a condition that could justify reimbursement for sex reassignment surgery. In recognition of the growing
cultural tolerance for a variety of gender-related identities, however, the *DSM-5* renamed sexual and gender identity disorders "gender dysphoria."[134] This new terminology implied that this condition was not truly a "disorder" despite its placement in psychiatry's classification of mental disorders. These dynamics illustrate
how *DSM* diagnostic decisions can depend far more on the particular interests and
values involved than on evidence-based standards.

The Revolution Will Not Occur

Just a year before the *DSM-5* was published, APA president-elect Jeffrey Lieberman proclaimed, "Since DSM-IV was released in 1994, there has been a wealth of
new research and knowledge about the nature and frequency of mental disorders,

how the brain functions and its neurobiology, and the lifelong influences of genes and environment on a person's health and behavior. Accordingly, the new edition—DSM-V—will reflect that new body of knowledge."[135]

In fact, the *DSM-5* closely resembles its predecessors: the grand ambitions of its creators are not evident in the diagnostic section of the manual but are found, if anywhere, in its appendices. It is difficult to see how the *DSM-5* is much improved, or even much changed, compared to its precursors. Evidence about brain functions, neurobiology, genes, and the like is nowhere to be found. Overall, the *DSM-5* maintains the descriptive, categorical, and theory-neutral approach of the three preceding manuals as well as their flaws. The search for etiologically informed diagnoses grounded in objective tests is no more advanced than it was when the first modern classifications emerged a century ago. External signs and symptoms are still the only resources available to categorize disorders.

The continuity of the *DSM-5* with its predecessors belied the extraordinary degree of hostility, controversy, and factionalism that marked its development. The leadership of the revision process and the NIMH had thoroughly repudiated and delegitimated the basic structure of previous *DSMs*. The APA membership and trustees, in turn, rejected researchers' aspirations to implement a new diagnostic paradigm. The process especially revealed the starkly different functions the *DSM* served for clinicians and researchers. Kupfer, Regier, and the task force, however, were tone deaf to the importance the *DSM* categories had assumed outside of the research community. Clinicians are often skeptical about the inherent reality of diagnoses but require them for practical reasons. One psychiatrist who combined both roles observed, "I am incredibly disciplined in the diagnostic classifications in my research, but in my private practice, I'll call a kid a zebra if it will get him the educational services I think he needs."[136] No single diagnostic system can satisfy both constituencies.

The DSM-5 Task Force failed to understand that psychiatric diagnoses cannot solely reflect the findings of research committees, epidemiological studies, and field trials. For most APA members, as well as for the institutions and other interest groups that rely on it, the *DSM* has become an indispensable practical tool. Clinicians were for the most part unconcerned with issues regarding dimensions, categories, or kappas. They did not have to believe in the validity of the *DSM* diagnoses, but they did have to put some code on a reimbursement form. Likewise, most patients and their families required a diagnosis to be reimbursed for treatment and receive other forms of valued resources. The extant *DSM* was well suited for their needs while a new dimensional system would have been a radical transformation with unknown consequences.

A dimensional system would have disrupted the many clinical, research, and administrative functions that depend on the *DSM*'s diagnostic system. Michael First summarized, "All mental health clinicians have been trained using the DSM categorical system. Psychology, psychiatry, and social work textbooks are organized using these diagnostic conceptualizations, practice guidelines have been developed based on these categories, and epidemiological data, service use and medical economic data, outcome data, and so forth have been compiled based on these categories."[137] The rank and file of the APA and outside advocacy groups was not willing to accept the uncertainties that a wholesale change of the *DSM* would have brought about. The embeddedness of diagnoses within clinical and institutional practice posed a formidable barrier to making any fundamental changes in the classification. The *DSM-5* will serve clinicians, administrators, families, and patients just as well as earlier *DSMs*.

The saga of the *DSM-5* was deeply unsettling for psychiatry. The participants in this process initially repudiated the *DSM* model but were thwarted in their determination to replace it. Although the manual has few, if any, fundamental changes from the previous versions, it emerged from the revision process with its credibility shaken. The myth that the *DSM* rested on a firm scientific basis was exposed by the same people who had propagated it. The *DSM-5* process might have signaled the beginning of the end of the manual's 40 years of unquestioned dominance in shaping the nature of mental disorder. Despite the overall similarity of the *DSM-5* to previous manuals, its critics were not satisfied. "This is the saddest moment in my 45 year career of studying, practicing, and teaching psychiatry," Allen Frances lamented at the time the *DSM-5* was approved. His blunt assessment was, "You don't buy it. You don't use it. You don't teach it." Even worse for the APA he urged, "It's important that the diagnostic system be taken away from the American Psychiatric Association."[138]

SEVEN | The *DSM* as a Social Creation

Since 1980 the *DSM* has fundamentally shaped nearly all aspects of psychiatric thought and practice. It overturned both the psychodynamic notion that mental disorders were intrinsically grounded in the fabric of each individual's existence and the social critique that madness was the consequence of civilization. The manual became the touchstone for the diagnoses that patients receive, that students are taught, and that researchers employ. Its labels dictate what mental health conditions insurance companies reimburse, government programs pay for, and drug companies market their products to treat. They also determine what mental illnesses are counted in the population, those that public policies address, and which ones shape community norms. Finally, *DSM* diagnoses frame how Americans and, increasingly, people worldwide understand their own interior problems.

Yet if the *DSM* powerfully impacts all facets of mental disorder, the manual itself results from the dynamics and organization of the psychiatric profession and wider cultural, political, and economic currents. Fluctuations in the politics of psychiatry, the nature of reimbursement for treatment, the marketing needs of drug companies, and the benefits patients, families, clinicians, and researchers receive from diagnoses shape the manual's uses. Over the course of the *DSM* period the manual changed from serving administrative purposes, to having little point at all, and then to becoming the starting point for understandings, explanations, prognoses, and treatments for mental disorder before the travails of the DSM-5 process challenged the manual's credibility. In short, the DSM is a fundamentally *social* document that both influences and reflects the changing internal and external dynamics surrounding psychiatry.

The Social Embeddedness of the DSM

The history of the DSM indicates the manual's deep entrenchment in the intraprofessional and general sociocultural forces that impact the psychiatric profession. Some of these influences arise from within the field. Between the mid-twentieth century and the present, psychiatric practice relocated from mental institutions to offices and clinics. At the same time, the profession's political center of gravity moved from the psychodynamic clinicians who dominated the initial DSM period to researchers, who held a very different diagnostic agenda. As well, changing medical norms forced the profession to adopt a more scientific-seeming classification system. Other forces came from outside the field. Once psychiatry moved from the hospital to the community, patients and their families were less likely to view diagnoses as imposed and unwanted labels than as valued sources of treatment and other resources. Federal agencies, schools, and insurance companies all made DSM conditions requirements for service provision. Drug advertisements became major channels for suggesting which diagnosis described a consumer's problems. All of these factors changed dramatically from the manual's emergence in 1952 through its most recent version in 2013.

INTRA-PROFESSIONAL FORCES

When the DSM arose, psychiatric practice was undergoing a thorough transformation. Before World War II nearly all psychiatrists were affiliated with mental hospitals, where diagnoses mostly served administrative purposes. The war radically altered the trajectory of the profession and its diagnostic system. It forced psychiatrists to confront the power of environmental stressors to produce mental disorders in previously normal individuals, dislodging the organic orientation that had shaped prior diagnostic manuals.

The DSM-I deftly responded to the new world of postwar psychiatry. It split mental disorders into organic and psychogenic categories, reflecting the recent division of hospital-based and community-oriented psychiatry. The profound psychic impacts of combat also led the initial DSM to add the far broader and more heterogeneous difficulties found in the general population to the largely psychotic problems that previous manuals emphasized. Shortly after the first *DSM* appeared, mental institutions began a precipitous decline. The roughly half million patients residing in state and county mental hospitals in the early 1950s fell to just 63,000 at the turn of the century and to below 40,000 at present.[1] The status of the major diagnostic categories changed accordingly. Compared to the initial duality of organic and psychogenic conditions, the most recent *DSM* devotes just 2 of its 22 chapters to the psychoses.

In contrast to institutions that had to classify their residents, outpatient practitioners had little need for diagnoses in the 1950s and 1960s. Most clients paid for their own treatment, so clinicians didn't require any *DSM* label to be reimbursed. Dominant analytic theories downplayed or dismissed the diagnostic process. Researchers were not yet an important force in the profession. Accordingly, the brief and nonspecific descriptions of the psychogenic conditions found in the first two *DSMs* sufficed.

In addition to the changing locus of psychiatric practice, political conflicts within the profession profoundly influenced the course of the *DSM*. The first *DSM* arose when almost all psychiatrists were clinicians or administrators. In contrast, by the early 1970s, psychiatric researchers had gained considerable power and prestige. For this group, diagnoses were essential starting points for their goals of uncovering the causes, courses, and outcomes for any condition. The extant *DSM* did not provide them with the precise descriptions and decision rules that ensured they were dealing with homogeneous groups. Robert Spitzer, the leader of the researchers who promulgated the *DSM-III* revolution, understood that the manual also had to meet clinical needs. Practitioners did not require uniform diagnostic criteria, and many held cynical attitudes about *DSM* diagnoses. By the 1970s, however, they had no choice but to put diagnostic codes on reimbursement forms. Clinicians also needed an inclusive system with categories that encompassed all their clients. Spitzer realized that the manual had to span the range of conditions found in psychiatric practice.

The former neuroses, now the anxiety and depressive classes, best illustrate how intra-professional concerns shape *DSM* diagnoses. In the initial two manuals, anxiety was the core dynamic behind all the psychoneuroses, including depressive neurosis. It was by far the most commonly found condition during the 1950s and 1960s. Unlike anxiety, depression was mostly associated with severe psychotic conditions and was much less likely to be diagnosed. In the following decade, while clinicians remained dedicated to the anxiety-based neuroses, the biological psychiatrists who were gaining power in the profession targeted depression.

The researcher-driven *DSM-III* inverted the relative standing of anxiety and depression diagnoses. Spitzer and his task force were determined to demolish the analytic concept of neuroses, which entailed removing anxiety from the core position it held in the previous two manuals. The *DSM-III* split anxiety into nine discrete conditions, which simultaneously decreased its sweeping scope. The carving up of the anxiety disorders contrasted with the unification of the formerly distinct psychotic and neurotic depressive conditions into a single major depressive disorder diagnosis. The *DSM-III* also allocated the most common symptoms of the neurasthenic tradition—for example, sadness, fatigue, sleep and appetite difficulties—to MDD rather than to any anxiety diagnosis. Further, the two-week duration it required for depression was far shorter than the typical six-month duration for anxiety conditions.

Intra-professional dynamics drove the divergent diagnostic criteria for anxiety and depression that arose in the *DSM-III*. Researchers were both attracted to depressive conditions and determined to sideline the anxiety-based neurotic diagnoses. Because its symptoms were so common and its criteria so easy to meet, MDD, rather than any anxiety disorder, became the diagnosis of choice in outpatient psychiatric and general medical practices. After 1980 the MDD diagnosis flourished. An unforeseen outcome of the *DSM*'s different treatment of depression and anxiety came when the SSRI drugs entered the market in the late 1980s. These drugs were not selective for depression, anxiety, or any other condition but act very generally to elevate levels of serotonin in the brain. The manual's distinct formulations of anxiety and depression led drug marketers to call them "antidepressants" rather than "anxiolytics." The enormous success of the "antidepressant" SSRIs echoed that of the "antianxiety" drugs of the 1950s and 1960s and reinforced the prominence of depression in the post-*DSM-III* period.

The intense intra-professional conflicts between clinical and research psychiatrists subsided during the *DSM-III-R* and *IV* era. Researchers embraced a diagnostic system that they expected would lead to major breakthroughs in understandings of and treatments for mental illnesses. Clinicians shed their former suspicions of a classification that brought them monetary rewards, elevated their status in the medical profession, and enhanced their cultural prestige.

The conflicts that rocked psychiatry during the making of the *DSM-III* re-emerged in the twenty-first century when researchers again attempted to revolutionize psychiatry's diagnostic system. By that time, the research community viewed the *DSM* categories, which had initially inspired such optimism, as obstacles to comprehending the nature of mental disorders. The new incarnation of the struggle between clinicians and researchers resurrected a question that has perennially bedeviled the profession: Are mental disorders best seen as categorical or dimensional? Diagnostic manuals before World War II generally relied on the categorical approach. In contrast, *Medical 203* and the first two *DSM*s, in line with Adolf Meyer's psychobiology and analytic views, moved in the direction of continuous conditions. Spitzer and his Washington University collaborators then returned to the earlier categorical model in developing the *DSM-III*. After the findings from neuroscientific research indicated a sharp disconnect from the *DSM* categories, the DSM-5 Task Force decided that the way forward lay in reverting to the conception of mental disorders as continuous.

The intra-professional politics during the *DSM-5* process highlighted the personality disorders. This class had been of central concern to dynamic clinicians during the *DSM-I and II* period. At that time, aside from the neuroses, they were the most frequently diagnosed conditions. In contrast, researchers scorned them because they did not fit the essential qualities of disease entities, resisted precise measurement, and featured vast interrelationships among their various types. The clash between clinicians and researchers over the inclusion of the personality disorders in the *DSM-III* was one of the most intense conflicts during the construction of that manual. Spitzer ingeniously mollified both researchers and clinicians by using the new multiaxial system to separate this class from the other *DSM* conditions.

In an interesting twist from prior *DSM* history, researchers, not clinicians, made the personality disorders central to their efforts to overhaul the *DSM* system. Allying with personality psychologists, the DSM-5 Task Force used this class as the model for its attempts to transform the manual. The Personality Disorders work group's proposal imported a common type of continuous measurement from psychologists, who studied populations where they had no need to make diagnoses or even consider actual behavior or real-life situations. Individuals receive a quantitative score on several personality dimensions that related their traits to the means of others in the population. Adherents to this school of thought would affix the prefix of "factor," not "psycho," before the term "analysis."[2]

If the establishment of the *DSM-III* signified the new power of psychiatric researchers, their failed attempt to transform the *DSM-IV* showed how deeply embedded in clinical practice the manual's diagnoses had become. This time, the intense conflicts over dimensionalizing the personality disorders were confined to the psychiatric profession. No drugs target this category, and no outside groups promote them. Earlier feminist opposition to diagnoses that targeted stereotypical female character traits had subsided. Even now, researchers continue to make strenuous efforts to dimensionalize the personality disorders, although whether they can impose a system that so radically departs from conventional clinical views in future *DSMs* is an open question.

When researchers involved in the *DSM-5* revision attempted to make fundamental changes in the manual, they found that the *DSM* categories had become so important for clinicians that they could not be dislodged. One reason was that practitioners worried that a dimensional system would threaten insurance reimbursement. The researchers who proposed a new dimensional system also underestimated the extent to which medical thinking and culture emphasizes dichotomies. While medical diagnoses are often uncertain and ambiguous, most diseases are distinct from—not continuous with—health. Even such dimensional conditions as blood pressure or cholesterol levels are divided at cut points that indicate likely pathology. Regardless of whether any illness is dichotomous or continuous in nature, clinicians must make decisions to treat or not to treat it. Therefore, the constraints of medical practice lead physicians, including psychiatrists, to think in black and white. Perhaps most important, diagnostic categories make mental disorders seem more *real* to the public, to physicians in other medical specialties, to insurance companies, and to federal regulators.[3] Any system that must legitimize its diagnoses as medical diseases will inevitably rely on categories rather than dimensions. In 1980 researchers successfully overcame clinical opposition to the new *DSM* system; by 2013 clinicians beat back the efforts of researchers to change the classification.

External Forces

Anxiety, depression, and the personality disorders illustrate how intra-professional disputes shaped some of the core *DSM* diagnoses. External interests played more important roles in developing a number of the manual's other conditions.

Before the early 1970s, no groups outside of the psychiatric profession showed much interest in the manual. At that time, the fierce public debate over homosexuality marked the first entry of outside forces into the *DSM* process. The success of LGBT organizations in removing this condition from the *DSM-II* turned out to be an exception. By the time the *DSM-III* appeared, unwanted diagnoses were already uncommon. Most subsequent attempts of patients, their families, and other external groups to change the manual involved broadening its existing conditions or incorporating new ones.

The determined efforts of Vietnam veterans to ensure that the *DSM-III* contained a PTSD diagnosis provided the model for future activists. Although its predecessor, gross stress reaction, stimulated the development of the *DSM-I*, the *DSM-II* lacked a comparable stress-related diagnosis. This gap did not concern the DSM-III Task Force, which believed that other categories such as anxiety and depression could capture the psychic impacts of environmental stressors. The lack of an appropriate stress-related diagnosis was, however, of intense concern to Vietnam War veterans, who launched a highly publicized, and ultimately successful, campaign to recognize persistent or long-delayed stress reactions as mental disorders. The veterans' advocacy efforts provided a model for lobbying that feminist groups employed to expand the PTSD criteria in the *DSM-III-R* and *DSM-IV*. The monetary benefits and access to care that can accompany a PTSD diagnosis have made this one of the *DSM*'s most popular diagnoses. "It is rare to find a psychiatric diagnosis that anyone likes to have," psychiatrist Nancy Andreasen observes, "but PTSD seems to be one of them."[4]

As third parties increasingly required *DSM* diagnoses to pay for treatment and other services, consumers and their families came to view them as valuable commodities. After federal legislation made the manual's conditions necessary tickets to educational and mental health benefits, as well as for monetary compensation through social security disability payments, the worth of the *DSM* labels was considerably enhanced. For example, by 2007, 1 in every 76 American adults received federal income assistance because of some mental disability.[5] Local and state governments also came to require *DSM* diagnoses before children could receive mental health treatments and special education services. Organized groups of parents helped drive the huge expansion of mental disorders among children and adolescents from 1980 to the present. Indeed, when changes in the *DSM-5* criteria sets for autism and Asperger syndrome threatened to restrict diagnoses, strong parental pushback forced the manual to allow all previously diagnosed children to maintain their labels even when they no longer met the new criteria. It is now far easier to broaden diagnostic criteria that lead to benefits than it is to reverse this process.[6]

As the use of the *DSM* turned from social control purposes to a necessary precondition for desired therapies, drugs, and other forms of assistance, the antidiagnostic narrative that was extremely influential during the 1960s and early 1970s became anachronistic. The *DSM* diagnoses were so highly valued that by the time the *DSM-5* revision began, even the LGBT organizations that led earlier conflicts to remove the homosexuality diagnosis from the *DSM-II* fought to maintain the gender dysphoria diagnosis because it was a precondition to pay for gender transition surgery. The *DSM-5* not only kept this condition at the insistence of activists but also gave it its own chapter, 1 of just 22 in the manual.

The pharmaceutical industry is another potent interest that has promoted the *DSM* diagnoses. Since 1972, the FDA had forced drug companies to demonstrate that their products were effective for specific mental disorders. Henceforth, all research on drug effectiveness would rely on *DSM* categories. In turn, psychiatry departments and many of their members became dependent on funding from pharmaceutical interests. Nearly 70 percent of DSM-5 Task Force members had financial ties to the industry.[7]

Since the 1990s, the powerful pharmaceutical industry has been inseparable from the rise of the *DSM* diagnoses into public consciousness. Drug companies were especially successful in capitalizing on parental desires to regulate their children's distressing and disruptive behavior. In addition to massive advertising campaigns, they generously funded advocacy groups concerned with childhood and adolescent conditions. Sales of stimulant drugs for ADHD quintupled from 2002 to 2013, by which time they amounted to almost $9 billion. Similarly, fueled by researchers who received generous funding from the drug industry, childhood bipolar disorder underwent a stunning fortyfold increase in rates from 1994 to 2003. The number of persons under 20 years old who received prescriptions for antipsychotic medications jumped from about 200,000 in 1993–95 to about 1.2 million in 2002.[8]

Over the course of the *DSM* era, the desirability of the manual's diagnoses have altered. When the *DSM-I* appeared, many psychiatric labels were imposed on involuntarily hospitalized patients or other socially deviant groups. By now, few are forced on resistant patients, but many bring about benefits for consumers and industries. *DSM* diagnoses lead to treatment for clients, services for families, reimbursement for clinicians, and profits for drug companies. Conversely, no powerful groups push any longer for abolishing *DSM* conditions. Both internal and external interests should ensure that the manual's large array of diagnoses persist in the years ahead.

Expanding Pathology?

Each *DSM* has had to grapple with the issue of how broad a range of pathology the manual should encompass. In one narrative, the hallmark of the *DSM*'s history lies in its increasing medicalization of emotions and behaviors. Medicalization is the reconceptualization of some previously nonmedical condition as a disease that health professionals ought to define and treat. This account emphasizes the growth of *DSM* diagnoses from 106 entities in the *DSM-I*, to 182 in the *DSM-II*, to 265 in the *DSM-III*, and to 292 in the *DSM-III-R*, before levelling off in the *DSM-IV* and 5. Much of this expansion involves medicalizing phenomena that had been seen as normal or as crimes, immorality, bad habits, or some other disreputable condition.

For some, "medicalization" is a descriptive, not an evaluative, term that does not judge whether the trend is desirable or undesirable. This group notes how diagnoses of mental disorder have expanded horizontally, to encompass new forms of pathology, and vertically, to capture milder forms of conditions that previously required greater severity.[9] Many critics, however, see growing medicalization as pernicious and strive to "protect normality from medicalization and psychiatry from overexpansion."[10] A third view, illustrated by anthropologist Roy Grinker's assessment of the growth of autism diagnoses, celebrates the benefits such as better access to treatment and services, lower levels of stigma, and improved research that this trend has brought about: "The prevalence of autism today is a virtue, maybe even a prize."[11]

One question the debate about medicalization raises is whether the range of behavior that falls under the legitimate authority of psychiatry has, in fact, increased. Undoubtedly, "the larger number of diagnostic categories in the DSM-III as compared with DSM-II reflects the clinical and research need for greater specificity in describing behavioral syndromes," as Spitzer observed.[12] It is less clear, however, that the scope of mental disorder itself has grown over the course of this period. The decades following World War II witnessed a tremendous broadening of behaviors that psychiatrists defined and treated.[13] The *DSM-I* and *II* diagnoses already encompassed much of the content of the more detailed criteria sets that arose in 1980. They were fewer in number but more expansive than the manuals that followed. Nevertheless, there is marked variation in the degree of newly medicalized conditions across the major classes of mental disorder.

Disorders among children and adolescents provide the clearest examples of how the *DSM* has increasingly medicalized previously nonmedical conditions. The initial *DSM* paid almost no attention to mental disorders among young people. It mentioned autism as a symptom of childhood schizophrenia and briefly described three categories of adjustment reactions in infancy, childhood, and adolescence. The *DSM-II* expanded this class, adding seven diagnoses to the original *DSM-I* conditions. Its category of behavior disorders of childhood and adolescence was 1 of just 11 major classes.[14]

The *DSM-III* greatly accelerated the enlargement of diagnoses among youth with a class renamed "disorders usually first evident in infancy, childhood, or adolescence." The manual placed this category first among its 15 general diagnostic classes. In contrast to the 3 pages the *DSM-II* devoted to this group, the *DSM-III* took 65 pages to describe 46 different criteria sets.[15] The following two revisions continued to broaden this class, especially for attention deficit / hyperactivity disorder, autism, and Asperger, a condition that first appeared in the *DSM-IV*. One result was that diagnoses of ADHD surged from less than 1 percent in 1980 to over 10 percent of children in 2011.[16] Another was that rates of all forms of autism expanded from just 1 in 2,000 children in the 1970s and 1980s to 1 in 68 when the *DSM-5* was published in 2013.[17] The *DSM-5* did not continue the growing medicalization of mental disorders among youth but moved many of the former age-defined diagnoses to their substantive homes (e.g., separation anxiety went to the anxiety disorders section; conduct disorder to disruptive, impulse, and conduct disorders). Its new class of neurodevelopmental disorders had significantly fewer categories than the child and adolescent class it replaced.

The bipolar II diagnosis that entered the *DSM* in 1994 provides another example of expanding pathology. This condition combined the easy-to-meet MDD criteria with a single hypomanic episode of at least four days' duration. A disorder that historically affected about 1 percent of the population came to encompass about five times that number.[18] It became the first *DSM* diagnosis to create a mass market for antipsychotic drugs.

The *DSM-5*'s expansion of the class of substance use disorder to encompass substance-related and addictive disorders provides a third instance of the extension of mental disorder into a new realm. The addition of an addictive category that is not substance related has huge potential consequences. While the *DSM-5* text incorporates only gambling disorders in the new category, the manual includes internet gaming disorder among its conditions for further study. Virtually all (~90 percent) Americans are now online, spending an average of between 18 and 24 hours a week on the internet.[19] This figure increases to about nine hours a day among teenagers.[20] This change potentially extends the notion of "addiction" from its traditional application to substances that induce compulsive consumption to virtually any activity that people frequently engage in, such as sex, shopping, eating, and exercising.[21] The *DSM*'s possible colonization of this huge new realm might realize Karl Menninger's belief that "most people have some degree of mental illness at some time, and many of them have a degree of mental illness most of the time."[22]

Medicalization displays a less straightforward trajectory for other major *DSM* classes. What the *DSM-I* called psychoneurotic disorders and the *DSM-II* neuroses had few specific categories but extremely wide reach. Outpatient psychiatrists and physicians could apply its vague formulations to the psychic consequences of poor marriages, economic worries, failed ambitions, and general nervousness. While the *DSM-III* and the manuals that followed it contained a far greater number of specific diagnoses, they did not expand the amount of pathology that the broad neurotic categories of the initial *DSM*s already captured.

The class of anxiety disorders also shows how a greater number of diagnoses need not indicate a growing medicalization of previously normal emotions. Although the *DSM-I* and *II* had fewer specific anxiety diagnoses than the manuals that followed, they provided a more expansive conception of this condition. The *DSM-II* conception was exceedingly broad: "Anxiety is the chief characteristic of the neuroses. It may be felt and expressed directly, or it may be controlled unconsciously and automatically by conversion, displacement and various other psychological mechanisms. Generally, these mechanisms produce symptoms experienced as subjective distress from which the patient desires relief."[23] The greater number of anxiety diagnoses in subsequent manuals did not expand pathology but more precisely defined conditions that earlier manuals had incorporated into their broad categories.[24]

The personality disorders are another case where recent *DSM*s are not more medicalized than were their predecessors. Although the particular types of conditions in this class changed considerably over the *DSM* era, neither their number nor their range increased. The first two *DSM*s presented capacious portrayals of 12 personality disorders that spanned from the most introverted to the most antagonistic as well as many character types in between.[25] Subsequent manuals added some conditions, abolished others, and changed the criteria for still others, but the *DSM-5* maintains diagnoses for 12 personality disorders.[26]

PTSD illustrates a third type of diagnostic trajectory that fluctuates in scope over the DSM period. Although the cultural presence of trauma-related diagnoses has expanded tremendously since 1980, gross stress reaction, growing out of the professional experiences of military psychiatrists in World War II, was a central diagnosis in the first *DSM*. Trauma-related diagnoses contracted in the *DSM-II* only to be resurrected in the *DSM-III*'s PTSD diagnosis. The broadening definitions of "trauma" in the PTSD criteria of the *DSM-III-R* and *DSM-IV* helped bring about an explosion of traumatic diagnoses.[27] The *DSM-5* changes to the PTSD diagnosis, however, constituted a rare example of an attempt to reduce the number of diagnoses through narrowing the scope of relevant traumas and limiting traumatic exposure to actual events.[28] The revised criteria set reversed the consistently growing expansion of PTSD from the *DSM-III* to the *DSM-III-R* and to the *DSM-IV.*

The development of the disruptive mood dysregulation disorder (DMDD) in the *DSM-5* is another unusual case of an attempt to reign in diagnostic expansion. The spectacular growth of pediatric bipolar disorders and resulting prescriptions for antipsychotic drugs was a huge embarrassment to the psychiatric profession in the early 2000s. The major proponents of PBD, Harvard psychiatrist Joseph Biederman and his associates, had received more than $4 million from Johnson & Johnson, the maker of the antipsychotic drug Risperdal. The resulting scandal generated widespread negative publicity, including front-page stories in the *New York Times* and a widely viewed segment on *60 Minutes* about a four-year-old who died from an overdose of drugs prescribed for her "bipolar disorder."[29] At the same time, the disruptive, oppositional, and irritable children who were diagnosed with PBD posed major behavior problems for their parents and others.

The *DSM-5* work groups on childhood and adolescent disorders developed an ingenious solution to the combination of the public repudiation of PBD and parental demands for some sort of treatment. After some missteps, they created a DMDD diagnosis that combined symptoms of irritable mood and aggressive behavior.[30] They placed it within the depressive disorders, not the bipolar and related disorders class, which severed its connection with the cycling of bipolar conditions that required powerful drugs. This placement signaled a departure from PBD and its attendant need for antipsychotic medication, yet preserved a diagnosis that would help parents deal with their troublesome children.[31]

A final category encompasses unsuccessful attempts to medicalize behaviors. The failed proposal for a psychosis risk syndrome in the *DSM-5* provides perhaps the most prominent example. As the previous chapter discussed, the PRS diagnosis strove to identify persons who did not meet criteria for a psychosis but who were thought to be at risk of developing one in the future. This attempt to greatly expand the realm of pathology met intense opposition, and the *DSM-5*, like its predecessors, lacks any diagnosis related to the risk of becoming mentally ill.

The depiction of an ever-growing realm of mental disorder that encompasses new forms of pathology and milder forms of previously recognized illnesses is, therefore, an overgeneralization. Childhood and adolescent disorders, the new class of addictive behaviors, and bipolar II are illustrations of a growing array of pathology to incorporate phenomena that had not previously been viewed as mental disorders. Yet, despite alterations in many of their particular criteria sets, the range of other diagnoses, such as the anxiety and personality disorders, has not incorporated previously non-pathological conditions. Some conditions, like PTSD, have both expanded and contracted over time. Finally, as PRS shows, some attempts at medicalization have failed. Almost no categories, however, have completely disappeared from the *DSM*, homosexuality being the rare exception.

Critics who object to the growing medicalization of previously nonmedical conditions argue that the expansion of mental disorder pathologizes normal experiences, stigmatizes the recipients of diagnoses, and generates unnecessary and often harmful treatments for people who don't need them. In addition, they note how treating non-disordered conditions takes resources away from those who genuinely could benefit from therapies.[32] These critics are unlikely to succeed. Regardless of their validity, these arguments are likely to hold little weight in the face of the substantial benefits that *DSM* diagnoses reap for many groups. Interests that support their maintenance or expansion include mental health professionals, who garner more clients; the drug industry, which gains a broader market; and patients and families, who get more services, desired treatments, and, often, government benefits. The professional, industry, and lay interests in preserving or increasing the number of diagnoses will likely continue to enfeeble objections to growing amounts of pathology.

On the Threshold?

Gerald Grob's synopsis concisely captures the cycles of advances and retreats that have marked the history of psychiatry:

> Every generation, moreover, insisted that the specialty stood on the threshold of fundamental breakthroughs that would revolutionize the ways in which mental disorders were understood and treated. In the nineteenth century the instrument of change was the mental hospital. In the mid-twentieth century, psychodynamic and psychoanalytic psychiatry became the vehicle by which the mysteries of normal and abnormal behavior would be revealed. At present the road to salvation is presumably through biological psychiatry, neuroscience, and genetics.[33]

As it enters the third decade of the twenty-first century, is psychiatry closer to revealing the mysteries of mental disorder than it was when the first *DSM* appeared?

In one prominent telling, the *DSM*'s evolution represents a tale of steady, if uneven, progress. The accumulation of new findings has led the manual to ever-better approximations of the reality of mental disorder. Prominent psychiatrists speak of how "scientific evidence" has replaced the charismatic authority of "great professors" in the evolution of the *DSM*.[34] The APA website lauds the DSM-5 Task Force: "Their dedication and hard work have yielded an authoritative volume that defines and classifies mental disorders in order to improve diagnoses, treatment, and research."[35] Indeed, some developments—the replacement of analytic assumptions with theory neutrality, the recognition that intense social stressors can produce lasting mental disorders, the removal of homosexuality, the acknowledgment of autistic disorders—do seem to improve the manual. Few psychiatrists would prefer the always-perfunctory and sometimes analytically infused definitions that prevailed before the *DSM-III* to the current *DSM-5* diagnoses.

Others paint a more skeptical picture of the history of the *DSM*. In their portrayals, the medical-like diagnoses that arose in the *DSM-III* are artifices that disguise psychiatry's continuing lack of progress in understanding the causes, prognoses, or optimal treatments for any of its major conditions. For example, sociologist Owen Whooley observes how every edition of the manual continues to camouflage psychiatry's fundamental ignorance about the basic nature of mental illness.[36] Historian Edward Shorter goes further, contending that the current classification is inferior to that of the original *DSMs*: "The diagnoses that flourished in the middle third of the twentieth century did a better job of cutting Nature at the joints than many of the diagnoses we have today, which are artifacts born of political compromises and sustained by pharmaceutical promotion rather than scientifically accurate descriptions of what is actually wrong with someone."[37] For such critics the major issues the *DSM* has had to confront over the past 70 years are no closer to resolution now than when the manual first arose.

Perhaps the basic barrier stymieing diagnostic progress is that—despite vast advances in brain-imaging technologies—psychiatry remains as reliant on observable symptoms as it was in the eighteenth century when Thomas Arnold wrote, "When the science of causes shall be complete we may then make them the basis of our classification, but till then we ought to content ourselves with an arrangement according to symptoms."[38] Like all psychiatric classifications that preceded it, the *DSM* system must still use reported symptoms, which can be organized in numerous ways, as the raw material for constructing its diagnoses. No attempt to develop etiologically informed diagnoses has yet to succeed. "Psychiatry is in the position—that most of medicine was in 200 years ago—of still having to define most of its disorders by their syndromes," eminent diagnosticians Robert Kendell and Assen Jablensky observe.[39]

The reliance on external symptoms especially hampers the construction of adequate definitions of the nature of mental disorder itself. One central, perhaps *the* central, issue regarding the *DSM* is what makes *any* of its diagnoses mental disorders. The *DSM-I* and *II* did not try to answer this question. Psychiatry was so well respected in the postwar period that no one questioned the field's authority to define its subject matter. This situation radically changed in the late 1960s and early 1970s when anti-psychiatry, gay, and feminist activists challenged the field's authority over what the term "mental disorder" should encompass. At the same time, strong cultural, political, and economic forces pushed psychiatry to conceive of its subject manner as disease entities comparable to those in the rest of medicine. Beginning with the *DSM-III* and persisting through the present, diagnoses reflect the idea that mental disorders, no less than physical diseases, have their own reality that is independent of particular life experiences. Extracting symptoms of mental disorder from the contexts in which they arise, however, is considerably more complicated than isolating symptoms is in the rest of medicine.

Chapter 3 described how Spitzer's experiences with attacks on the homosexuality diagnosis and with confronting the sweeping anti-psychiatry critique that the field could not define even its central concepts led him to recognize that the *DSM-III* required some general statement that distinguished appropriate from inappropriate uses of the term "mental disorder." A major obstacle he faced was that a core principle of the *DSM-III* revolution was to use observable symptoms without regard to their underlying causal mechanisms to define each diagnosis. The turn to classifying mental disorders by their outward appearances hindered the ability to separate mental disorders from symptomatically similar but contextually explicable responses and from disliked and devalued but not disordered responses.

The absence of a definition of "mental disorder" did not bother many psychiatrists, who noted that no medical specialty defined what an organic disease is. Why, they asked, should psychiatry be different from other branches of medicine? This objection persists at present. "When someone has a myocardial infarction (MI), physicians regard it as an instantiation of cardiac disease, regardless of its 'context,'" psychiatrist Ronald Pies claims. Further, "the MI may have occurred in the context of the patient's poor diet, smoking, and high levels of psychic stress—but it is still an expression of disease."[40] For Pies and others, just as a heart attack is a heart attack, a mental illness is a mental illness. Therefore, like the rest of medicine, context is irrelevant to separating psychiatrically disordered symptoms from situationally apt or culturally appropriate expressions.

Yet Spitzer realized that defining "disorder" is fundamentally different—and far more challenging—in psychiatry than in physical medicine. Separating organic symptoms from the settings in which they occur was a hallmark of diagnostic progress in other medical specialties.[41] In psychiatry, however, divorcing symptoms from context has the opposite impact of hopelessly blurring situationally appropriate psychological phenomena from mental disorders. This is because all mental functions are highly sensitive to environmental circumstances. Virtually every symptom of various mental disorders can sometimes be biologically and psychologically suitable adaptations to given contexts, culturally explicable expressions, or both. For example, symptoms resembling depression that arise after the death of a loved one indicate that grief mechanisms are working appropriately, not inappropriately. Likewise, a panic attack is an understandable response when facing an impending fall off a cliff but a sign of disorder in the absence of danger.[42] Or hearing voices, which can be a hallmark of schizophrenia, is sometimes explicable in particular cultural and religious settings.[43] In contrast, a heart attack always signals a failure of natural functioning regardless of the context or culture in which it emerges. Unlike other medical specialties, context is an intrinsic aspect of deciding what a mental disorder is or is not.[44]

Spitzer understood the difficulties that developing a general definition of "mental disorder" entailed but nevertheless realized such a statement was necessary to establish the field's credibility and to protect it from anti-psychiatry attacks. His original formulation, presented in chapter 3, has mostly endured through the present *DSM-5* version:

> A mental disorder is a syndrome characterized by clinically significant disturbance in an individual's cognition, emotion regulation, or behavior that reflects a dysfunction in the psychological, biological, or developmental processes underlying mental functioning. Mental disorders are usually associated with significant distress or disability in social, occupational, or other important activities. An expectable or culturally approved response to a common stressor or loss, such as the death of a loved one, is not a mental disorder. Socially deviant behavior (e.g., political, religious, or sexual) and conflicts that are primarily between the individual and society are not mental disorders unless the deviance or conflict results from a dysfunction in the individual, as described above.[45]

This characterization both expresses what mental disorders are—dysfunctions of some mental process that is not working as it should and that lead to distress or disability—and what they are not—culturally defined deviant behaviors or conflicts between individuals and society. The DSM's succinct definition identifies both essential aspects of mental disorders and separates them from states that are often mislabeled as psychiatric problems.[46]

Spitzer knew that the definition couldn't set an exact boundary between dysfunctions and normal-range behaviors because no sharp division exists in nature. The borders between mental disorders and non-disordered states of distress or social deviance are often fuzzy, vague, and ambiguous. Despite this caveat, the statement serviceably separates distressing states that are proportionate responses people make to stressful social arrangements or are norm-violating behaviors from mental disorders that arise because of dysfunctional psychological mechanisms. When the particular criteria sets for each DSM diagnosis take this general definition into account, they should strike the best balance between recognizing true cases of disorder and weeding out contextually appropriate or socially devalued behaviors.[47]

The central problem has been that, in practice, many of the particular DSM criteria sets don't follow the definition.[48] Worse, criteria for what makes some core diagnoses mental disorders have deteriorated, rather than progressed, over time. Depression provides probably the most egregious example; succeeding DSMs have increasingly relied on external symptoms without regard to the context in which they arise. The first DSM directed clinicians to consider "the realistic circumstances of the loss," and the DSM-II treated only "excessive" symptoms as signs of a disorder.[49] The three DSMs that followed provided far more symptom-based criteria sets for MDD but at least excluded uncomplicated symptoms that arose from bereavement from diagnosis. The DSM-5 criteria set eliminated even this exception.[50] Because depressive symptoms need last just two weeks, they can easily be short-lived responses to events rather than true disorders. Far from being "on the threshold" of growing understandings, the context-free DSM-5 MDD diagnosis shows even greater heterogeneity and indistinct boundaries with normal sadness than did its predecessors.

In other cases, DSM criteria do not separate mental disorders from social deviance. The history of substance use disorder (SUD) diagnoses recounted in the previous chapter provides the primary example of this type of growing conflation. The first two DSMs placed most SUD diagnoses among conditions that involved brain damage. The DSM-III greatly broadened this category by adding to addictive behaviors a new abuse diagnosis that referred to the problematic social consequences of substance use. Abuse and addiction remained separate diagnoses in subsequent revisions until the *DSM-5* combined them into a single category, in defiance of the admonition in the manual's definition of mental disorder to avoid mislabeling social deviance as dysfunction. Instead, the new criteria for SUD blur dysfunctions that stem from addictions with rule violations and social impairments.

In other cases, diagnoses do not distinguish dysfunctions from conflicts between individuals and society. The new diagnosis of hoarding disorder in the *DSM-5* illustrates this situation. Its essence is "persistent difficulties discarding or parting with possessions regardless of their actual value."[51] Hoarders typically accumulate so many items that they or others have difficulty navigating their living spaces. This telegenic situation has become the object of a popular television series, *Hoarders*. Yet the new diagnosis potentially involves many false-positive problems: hoarders themselves are typically not troubled by their condition unless someone tries to *stop* their stockpiling. Interventions typically arise after family members, neighbors, or public health departments complain. The justification for calling hoarding behaviors "dysfunctions" or "mental disorders" as opposed to "conflicts between individuals and society" is unclear.[52]

Although the overall narrative of growing diagnostic progress over the *DSM* era is, at best, questionable, the *DSM-5* did take some steps to better distinguish mental disorders from other conditions. One was its replacement of the former little-used Axis IV, psychosocial and environmental problems, with a new chapter on other conditions that may be a focus of clinical attention. The *DSM-5* forthrightly states, "The conditions and problems listed in this chapter are not mental disorders."[53] They encompass relational difficulties; abuse and neglect; and educational, occupational, housing, legal, and economic problems. The inclusion of a wide range of psychosocial, personal, and environmental difficulties that are explicitly not mental disorders could help clarify the hazy boundaries between mental disorders and contextually appropriate distress that characterized previous manuals.

The *DSM-5* also took steps to incorporate cultural differences in symptomatic expressions into account. It includes a whole chapter on cultural formulation based on the idea that "understanding the cultural context of illness experience is essential for effective diagnostic assessment and clinical management." Many of its particular criteria sets note these differences. For example, GAD symptoms can present somatically in some cultures and cognitively in others. In schizophrenia, "ideas that appear to be delusional in one culture (e.g., witchcraft) may be commonly held in another. In some cultures, visual or auditory hallucinations with a religious content (e.g., hearing God's voice) are a normal part of religious experience."[54] Overall, the latest manual is more attentive to errors that arise from mistaking cultural differences with mental disorders. Still, future DSMs will continue to face challenges in distinguishing psychiatric dysfunctions from contextually and culturally explicable responses and from social deviance.

Future DSMs

The evolution of the DSM reveals the huge, and possibly unresolvable, difficulties in defining mental disorders. At first, the manual's developers bet that combining organic conditions found in asylums with a psychodynamic model that captured the problems of outpatients was sufficient to meet psychiatry's diagnostic needs. Next, they embraced a medical model that assumed a basic resemblance between mental and physical disorders but refrained from theoretical speculations about their causes. Once it became clear that no etiological basis was on the horizon, the DSM-5 unsuccessfully tried to adopt the statistical techniques and dimensional measurements of psychological research to define its conditions. The latest, as yet totally unrealized, efforts assume that neuroscientific research will point the manual in a more brain-based direction. Whether the research domain criteria, the NIMH's ambitious new diagnostic system grounded in neural circuitry, will succeed is far from assured at present.

The current DSM contains 22 general classes and nearly 300 specific diagnoses. Yet perhaps the most striking finding from neuroscientific research is that the hundreds of DSM diagnoses reflect variations on a small number of general processes that are loosely related to internalized, externalized, and psychotic disorders.[55] A DSM organized around just three classes, however, would be professionally unthinkable: such a system would have no medical authority. The lack of alternatives to using external symptoms as the basis for diagnosis ensures that future DSMs are unlikely to result in any fundamental breakthroughs in understanding and treating mental disorders.

The DSM-5 process sunk the credibility of the manual to levels not seen since the 1970s. Critics of the latest revision were not the vociferous anti-psychiatrists who objected to earlier versions but eminent figures within the profession, including Spitzer and Frances, the chairs of earlier DSM revisions, as well as former NIMH directors Steven Hyman and Thomas Insel. Most perversely of all, the leaders of the DSM-5 Task Force themselves vigorously critiqued the manual. Only after their efforts at paradigm change failed did they revert to defending the extant diagnostic system. Ironically, the major guardians of the classification were clinicians who were just as skeptical as researchers about the DSM's validity but required its diagnoses for practical purposes. In the face of such widespread recognition of the manual's weaknesses, will it be possible to resurrect the credence of the *DSM* enterprise?

The *DSM-III* era when research needs were tied to categorical diagnoses seems to be over: researchers no longer respect the symptom-based *DSM* entities and are searching for neuroscientific alternatives such as the RDoC. From the opposite direction, calls arise for a return to the era when psychiatry focused on personal history and interpersonal connections. One group of British psychiatrists calls for junking the *DSM* system because "good practice in psychiatry primarily involves engagement with the non-technical dimensions of our work such as relationships, meanings and values."[56]

Despite continuing frustrations over establishing a valid diagnostic classification, there is little chance of replacing the *DSM* anytime soon. Patients require its diagnoses to obtain treatment, clinicians to get reimbursement, insurance companies to make payments, schools to provide services, government programs to decide eligibility, and drug companies to market products. Perhaps most important, the current *DSM* fulfills psychiatry's need for professional legitimacy; all medical specialties require specific diagnoses. The *DSM* system will persist into the foreseeable future because its diagnostic entities are closely intertwined with too many interests that find it impossible to give them up.

Observable symptoms continue to define the *DSM* diagnoses, a situation that general medicine surmounted more than a century ago. The chemical and physical operations of the brain have yet to, and might never, provide clues that may unravel the mysteries of human consciousness and its distortions. Unlike their physical counterparts, mental disorders could require understandings that cannot be completely removed from personal life experiences. The future of the *DSM* is clearly at a crossroads, but the path it should take has no roadmap.

Notes

CHAPTER ONE: **Diagnosing Mental Illness**

1. Havermann, 1957, 72.
2. In Staub, 2011, 38.
3. Advertisement for Ultran in Smith, 1985, 103.
4. Rosenberg, 2007, 13.
5. APA, 1980, 6.
6. Horwitz & Wakefield, 2012, 23.
7. World Health Organization, 2018; Lane, 2007, 3.
8. E.g., Smith, 2019.
9. Gorenstein, 2013.
10. Slovenko, 2011.
11. Kessler et al., 2005.
12. Caspi & Moffitt, 2018.
13. E.g., Leaf, Myers, & McEvoy, 1991; Kessler et al., 2005; Hasin & Grant, 2015.
14. Lieberman, 2015, 87.
15. Sabshin, 1990, 1272; Maxmen, 1985, 31; Lieberman, 2015, 292.
16. Scull, 2019, 54.
17. Szasz, 1961.
18. Bracken et al., 2012, 434.
19. Caplan, 1995.
20. E.g., Foucault, 1965; Staub, 2011.
21. Frances, 2012b.
22. Rosenberg, 2007, 32. Polish microbiologist Ludwig Fleck developed the notion of "thought collective" in his 1935 book, *Genesis and Development of a Scientific Fact*. See Zerubavel, 1991, 1999, for a general view of the social perspective.
23. John Charles Bucknill and Daniel Hack Tuke quoted in Shorter, 2015, 160.
24. Kirk & Kutchins, 1992, 222.
25. Pilecki, Clegg, & McKay, 2011.
26. Dan Pine quoted in Grinker, 2007, 135.
27. Verhoeff, 2010, 469.
28. Although conditions that would meet current diagnostic criteria for schizophrenia did not arise until a much later period, ancient descriptions portrayed persons with delusions and hallucinations that are the most characteristic features of schizophrenia. Such people have always suffered more rejection, stigma, and social fear than has any other diagnostic group.

29. See especially Shorter, 2013.
30. Horwitz, 2018.
31. Lunbeck, 2014.
32. Edwards, 2012, 699.
33. This book doesn't discuss the *DSM-IV-TR* (2000), which was for the most part confined to cosmetic revisions of the *DSM-IV* (1994).

CHAPTER TWO: **The *DSM-I* and *DSM-II***

1. See Simon, 1978; and Jackson, 1986, for good descriptions of Hippocratic conceptions of mental illness.
2. Dain, 1964, 16.
3. Wallace, 1994, 62. There were exceptions. Most notably, Scottish physician William Cullen developed a classification that contained hundreds of types of mental disorders (Horwitz, 2013, 51–52).
4. Arnold quoted in Lewis, 1967, 193.
5. Weiner, 1994.
6. This section is adapted from Grob & Horwitz, 2010, 25–27.
7. Dain, 1964, 72.
8. Pliny Earle to Clark Bell, April 16, 1886, Earle Papers, American Antiquarian Society, Worcester, MA.
9. Noll, 2011, 26.
10. Ray, 1838/1962, 59–60.
11. This schema was also used in the census of 1890 but was discontinued thereafter (Kramer, 1968, xi–xii).
12. Tuke, 1892, 229, quoted in Lewis, 1979, 192.
13. Beard, 1869.
14. See especially Ropper & Burrell, 2019.
15. Ropper & Burrell, 2019, 113.
16. Noll, 2011, 63, 66.
17. Kraepelin, 1896, quoted in Decker, 2013, 45.
18. Paranoids were marked by persecutory delusions, catatonics by extreme activity variations from very inhibited to very agitated, and hebephrenics by unusual behavior and moods.
19. Kraepelin, 1899, quoted in Shorter, 2015, 76–77.
20. Shorter, 2013, 100. By the end of his life, Kraepelin even questioned the basic division of dementia praecox and manic depression (Noll, 2011, 259).
21. Bleuler, 1911.
22. Tone, 2009, 18.
23. Berkley, 1900, 97–98; Paton, 1905, 225–27.
24. Raines, 1952, v.
25. Hill, 1907.
26. Horwitz, 2013, 81–84.
27. Pols, 2001.
28. Before 1938, patients did not require a physician's prescription to obtain any medicine (Herzberg, 2009, 22).

29. E.g., Pressman, 2002; Scull, 2005.

30. Wallace, 1994, 75–76; Noll, 2011, 47.

31. Menninger, 1963, 466.

32. *Statistical manual*, 1918.

33. Grob & Horwitz, 2010, 27. Although Meyer had enthusiastically promoted Kraepelin's ideas earlier in his career, over time he came to sharply turn against them.

34. Grob, 1990, 53.

35. Raines, 1952, vi–vii.

36. Grob, 1994, 197.

37. Menninger, 1945; Houts, 2000.

38. War Department, 1946.

39. War Department, 1946, 467. The manual listed six types of schizophrenic reactions: latent, simple, hebephrenic, catatonic, paranoid, and unclassified. It divided paranoid disorders into paranoia and paranoid states. Finally, it listed manic depression, psychotic depression, and involutional depression as the three types of affective disorders.

40. Raines, 1952, vii.

41. Grob, 1991a, 42.

42. Decker, 2013, 7.

43. Rickles, Klein, & Bassan, 1950.

44. E.g., Meyer, 1948.

45. Meyer quoted in Noll, 2011, 160.

46. Grinker, 1965, quoted in Whooley, 2019, 120.

47. Alexander, 1956, quoted in Whooley, 2019, 109.

48. GAP, 1950, quoted in Grob, 1994, 199.

49. Herman, 1995, 249.

50. Quoted in Grob, 1991a, 429.

51. By 1962 well over half of the heads of psychiatry departments were members of psychoanalytic organizations (Hale, 1995, 253).

52. Grob, 1994, 199–202, 218.

53. At the time, this manual was simply called *DSM*. Its promulgators, of course, had no way of knowing that future additions would add roman numerals so that the initial manual would become the *DSM-I* after 1968.

54. Kramer, 1953.

55. Grob, 1991a.

56. Menninger, 1963, 9.

57. APA, 1952, 12.

58. APA, 1952, 14.

59. The manual also contains a separate class of mental deficiency that was "primarily a defect of intelligence existing since birth, without demonstrated organic brain disease or known prenatal cause" (APA, 1952, 23–24).

60. The full name was disorders of psychogenic origin or without clearly defined physical cause or structural change in the brain (APA, 1952, 24).

61. Houts, 2000.

62. Although American psychiatry generally accepted Kraepelin's term "dementia praecox" early in the twentieth century, over the course of the century it increasingly deployed Swiss psychiatrist's Eugen Bleuler's concept of "schizophrenia." The field used these terms interchangeably through the 1920s and part of the 1930s until "dementia praecox" dropped out of the field's lexicon by 1938 (Noll, 2011, 233, 262).

63. Hale, 1995, 209.

64. APA, 1952, 29.

65. APA, 1952, 12

66. APA, 1952, 33–34.

67. Shorter, 2009, 158.

68. APA, 1952, 25.

69. APA, 1952, 34.

70. Healy, 1997, 235.

71. APA, 1952, 34–35.

72. APA, 1952, 38.

73. APA, 1952, 39.

74. APA, 1952, 40.

75. Sadler, 2005, 151.

76. Hale, 1995.

77. Menninger, 1963, 478.

78. Raines, 1953.

79. Hunter & Macalpine quoted in Porter & Micale, 1994, 10.

80. Despite the intention to reconcile the *DSM* with the *ICD*, the *DSM-II* contained many differences from the *ICD* (Skodol, 2000, 432).

81. Spitzer quoted in Wilson, 1993, 406.

82. Mayes & Horwitz, 2005, 251.

83. APA, 1968, 48.

84. APA, 1968, viii.

85. APA, 1968, 39, 40.

86. Decker, 2013, 95.

87. The 10 categories were mental retardation, organic brain syndromes, psychoses not attributed to physical conditions, neuroses, personality disorders, psychophysiologic disorders, special symptoms, transient situational disturbances, behavior disorders of childhood and adolescence, and conditions without manifest psychiatric disorder.

88. Healy, 1997, 251.

89. Fine, 1979, 148.

90. Porter & Micale, 1994, 13.

91. For example, Holden Caulfield in the *Catcher in the Rye* sees a psychiatrist; the best-known poem of the era, Allen Ginsberg's *Howl*, features his mother's insanity; *Catch-22* refers to the use of psychiatric excuses.

92. O. Hobart Mowrer, quoted in Burnham, 2012, 158.

93. Cooper & Blashfield, 2016, 454.

94. Berger quoted in Shorter, 2009, 150.

95. National Disease and Therapeutic Index, 2019. I am grateful to David Herzberg for providing me with these data. I am solely responsible for their interpretation.

96. Herzberg, 2009, 260n26. In 1968 total office, hospital, and nursing home visits were 20,862,000 psychoneuroses and disorders of personality, 11,886,000 anxiety reaction, 8,793,000 psychoses, 7,694,000 neurotic depression, 2,542,000 alcoholism, and 555,000 mental deficiency (*NDTI Review*, 1970, 22).

97. Kadushin, 1969, 103. See also Rieff, 1966.

98. Schnittker, 2017, 182–83.

99. Tone, 2009, 90; Smith, 1985, 67.

100. Metzl, 2003, 73; Tone, 2009, xvi; Smith, 1985, 27; Herzberg, 2009, 30; Tone, 2009, 27, 90; Healy, 1997, 65.

101. Herzberg, 2009, 34.

102. Smith, 1985, 102, 103.

103. Herzberg, 2009, 35.

104. Metzl, 2003, 72, 103. During the 1950s the minor tranquilizers were promoted for men and women alike. For example, an ad used in medical journals indicated that tranquilizers could be used in situations such as "a lawyer involved in preparing a complicated legal brief or an overworked executive carrying a tremendous burden of responsibility." In the 1960s, however, their appeals were primarily directed toward women. Unlike Miltown, which was equally targeted to both men and women, the benzodiazepines were clearly aimed at women's problems. Women obtained about twice the number of prescriptions for these drugs as men did (Herzberg, 2009, 37, 48). Ads rarely, however, invoked dilemmas that working-class or poor people faced. Virtually all promotions featured middle- or upper-middle-class problems.

105. Herzberg, 2011, 418.

106. Whiteside, 1958, 117.

107. Friedan, 1963/2001, 293.

108. Gardner, 1971.

109. Mead quoted in Smith, 1985, 181.

110. Shorter, 2009, 99; Herzberg, 2009, 138; Herzberg, 2011, 419.

111. This issue devoted 108 pages to substantive content.

112. Tone, 2009, 155; Smith, 1985, 113, 119–20; Herzberg, 2009, 193.

113. Blackwell, 1975; Smith, 1985, 37; Herzberg, 2009, 40.

114. Herzberg, 2009, 18, 30.

115. Kirk & Kutchins, 1992, 202.

116. My account is especially indebted to Bayer, 1987.

117. McNally, 2011, 21–22.

118. APA, 1968, 44. The Feighner criteria that research psychiatrists published in 1972 and were to become the model for the *DSM-III* listed homosexuality—"persistent homosexual experiences beyond age 18"—as 1 of just 14 diagnoses (Feighner et al., 1972, 61).

119. Psychoanalysts, in particular, were astonished that a standard such as "subjective distress" could establish whether a condition was a mental disorder. They were also amazed that outside groups could have such powerful influences over what they considered to be a purely medical decision. Disgruntled analysts sponsored a subsequent referendum to overturn the decision, which was defeated by a margin of 58 to 37 percent (Bayer, 1987, 148).

120. Bayer, 1987, 138.

121. APA, 1974, vi.

122. Spiegel, 2005; and Decker, 2013, provide good biographical material about Spitzer.

123. Spitzer & Wilson, 1968.

CHAPTER THREE: **The Path to a Diagnostic Revolution**

1. Skodol, 2000, 432.

2. Klein's work showed that imipramine led panic attacks but not generalized anxiety disorder (GAD) to cease. This seemed to indicate that panic and GAD were two distinct disorders (Klein & Fink, 1962). See also Callard, 2016. Later studies, which indicated that GAD also responded to imipramine, called this finding into question. See Rickels & Rynn, 2001.

3. Pollock, 1959, 314.

4. Metzl, 2003, 104–5; Healy, 1997, 66.

5. Klerman, 1983, in Healy, 1997, 247.

6. Hale, 1995, 246; Grob, 2011, 413.

7. Grob & Goldman, 2006.

8. Healy, 1997, 100.

9. Smith, 1986, 81.

10. Edwards quoted in Smith, 1986, 188, 127, 189.

11. Smith, 1986, 217, 195, 210, 49, 226.

12. Shorter, 2009, 116.

13. Healy, 1997, 163.

14. See Gabbard & Gabbard, 1999.

15. Szasz, 1961; Rosenhan, 1973.

16. Rosenhan, 1973, 251.

17. Hale, 1995, 341.

18. Veroff, Kulka, & Douvan, 1981, 79.

19. Murray, 1979, 255; *NDTI Review*, 1970, 2, 1–5; Kramer, 1977; Grob & Goldman, 2006, 50.

20. Hackett, 1977, 434.

21. Klein quoted in Spiegel, 2005, 58.

22. Spitzer to Gerald C. Davison, April 15, 1976, Archives, DSM Coll.

23. Wilson, 1993, 408.

24. Phyllis Greenacre quoted in Hale, 1995, 307.

25. Menninger, 1963, 325.

26. APA, 1980, 6.

27. Spitzer quoted in Decker, 2013, 147.

28. Harvey Bluestone, Hector Jaso, & Howard E. Berk to DSM-III Task Force, September 30, 1976, Archives DSM Coll.; Baltimore/DC Society of Psychoanalysis to H. Keith Brodie, May 10, 1979, Archives, DSM Coll.

29. Feighner et al., 1972. The diagnoses were the primary affective disorders of depression and mania, secondary affective disorders associated with some other psychiatric or medical diagnosis, schizophrenia, anxiety neurosis, phobic neurosis, hysteria,

antisocial personality disorder, alcoholism, drug dependence, mental retardation, homosexuality, transsexuality, organic brain syndrome, and anorexia nervosa, as well as a residual category of undiagnosed disorder.

30. Kendler, Munoz, & Murphy, 2010.

31. Spitzer, Endicott, & Robins. 1978, 781.

32. Lawrence Rockland quoted in Bayer & Spitzer, 1985, 191.

33. The *ICD-9* was eventually published in 1979.

34. Bayer & Spitzer, 1985, 187.

35. Spitzer quoted in Spiegel, 2005, 63.

36. Spitzer quoted in Spiegel, 2005, 59.

37. Decker, 2013, 108.

38. See especially Decker, 2013, 234, 238.

39. Meehl, 1972, 79.

40. Spitzer, Endicott, & Robins, 1975, 1188.

41. APA, 1968, ix.

42. Beck et al., 1962; Pasamanick, Dinitz, & Lefton, 1959, 127.

43. Kendell et al., 1971. In 1972, eminent diagnostician Paul Meehl (1972, 273) expressed a contrary view: "It is not true that formal nosological diagnosis is as unreliable as the usual statements suggest." Meehl believed that studies confined to major diagnostic categories such as schizophrenia, adequate clinical exposure to patients, and well-trained clinicians would produce reliable results.

44. Kirk & Kutchins, 1992, 64, 261; Spitzer & Williams, 1983; Spitzer & Fleiss, 1974.

45. Frances quoted in Spiegel, 2005, 58.

46. Talbott, 1980, 27.

47. Kirk & Kutchins, 1992, 185.

48. Millon quoted in Angell, 2009, 29.

49. Horwitz & Wakefield, 2007, 88–89.

50. Lewis, 1934.

51. Kiloh & Garside, 1963.

52. Overall et al., 1966; Hamilton & White, 1959; Paykel, 1971; Raskin & Crook, 1976.

53. Kiloh et al., 1972; Everitt, Gourlay, & Kendell, 1971.

54. Davies, 2017, 38.

55. Guze quoted in Healy, 2000, 407.

56. Millon quoted in Kirk & Kutchins, 1992, 203.

57. Spitzer quoted in Shorter, 2009, 157.

58. Maxmen, 1985, 83.

59. APA, 1980, 7. Not only was the *DSM-III* theory neutral, but also it did not dictate any form of treatment. The purpose of the manual was to describe mental disorders, not to point to particular therapies for any of its conditions. Its introduction observed that all clinicians—psychodynamic, behavioral, somatic, and family oriented—could use its many diagnoses.

60. Roger Peele to Spitzer, March 12, 1979, Archives, DSM Coll.

61. Kernberg quoted in Bayer & Spitzer, 1985, 190.

62. Bayer & Spitzer, 1985.

63. *ICD-9*, code 300. A comparative analysis of the *DSM-III* and the *ICD-9* commissioned by the APA shortly after the *DSM* was published concluded that "there is a serious degree of incompatibility between the two classifications for mental disorders." Robert H. Seeman, "A comparative analysis of ICD-9-CM and DSM-III" to American Psychiatric Association, October 10, 1980, Archives, DSM Coll. The Public Health Service advised the APA that it would not adopt the *DSM-III* until these incompatibilities were resolved. Robert A. Israel to Darrel A. Regier, January 6, 1981, Archives, DSM Coll.

64. Bayer & Spitzer, 1985, 192.

65. Hector Jaso and Howard E. Berk to Spitzer, June 29, 1976, Archives, DSM Coll. "Levittown" referred to a group of suburban housing developments with mass-produced identical dwelling structures.

66. Klein quoted in Wilson, 1993, 407.

67. Bayer & Spitzer, 1985, 191.

68. Lane, 2007, 26.

69. Bayer & Spitzer, 1985, 195.

70. Henry Pinsker to Task Force on Nomenclature and Statistics, June 4, 1975, Archives, DSM Coll.

71. John K. Wing to Spitzer, November 5, 1976, Archives, DSM Coll.

72. Spitzer knew that few others shared his passion to develop a general definition of mental disorder. He wrote to one psychiatrist who responded favorably to his efforts: "Anyone who finds my attempt to define mental disorder useful, must be a very special person." Spitzer to Lyman C. Wynne, July 19, 1976, Archives, DSM Coll.

73. Szasz, 1961.

74. Spitzer (1975) noted how most of these patients were diagnosed with "schizophrenia in remission," a diagnosis that was extremely rare among genuine patients but an accurate characterization of the behavior of Rosenhan's pseudopatients. Cahalan (2019, 193) notes that Spitzer considered his critique "the best thing I have ever written." Her book indicates that the Rosenhan study was largely a fabrication. Perhaps most interesting of all, Cahalan reveals that Spitzer knew that the study was mostly Rosenhan's own invention. He did not disclose this because he was deeply invested in making the same point as Rosenhan: the extant diagnostic system had no credibility.

75. Leon North quoted in Decker, 2013, 157.

76. Spitzer, Sheehy, & Endicott, 1977, 4. See also Spitzer to Martin Roth, June 3, 1976, Archives, DSM Coll.; John E. Cooper to Spitzer, November 5, 1976, Archives, DSM Coll.

77. Theodore H. Blau to Robert Gibson, August 8, 1977, Archives, DSM Coll. Spitzer told Melvin Sabshin that he would "enjoy" making Blau's letter public so that psychiatrists would realize that Spitzer was making every possible effort to define psychiatry as a medical discipline. Spitzer to Sabshin, October 26, 1977, Archives, DSM Coll.

78. Jack Wineberg to Theodore H. Blau November 3, 1977, Archives, DSM Coll.

79. Theodore H. Blau to Jack Wineberg, Dec. 6, 1977, Archives, DSM Coll.

80. APA, 1980, 6.

81. Talbott, 1980, 26. Spitzer (1999) later adopted philosopher Jerome Wakefield's somewhat more specific notion that a dysfunction refers to some psychological mechanism that is not working as evolutionary processes designed it to work.

82. E.g., Clark, 1999.

CHAPTER FOUR: **The *DSM-III***

1. Whooley, 2019, 175.
2. Klerman, 1984, 25.
3. Nearly all commentators assume the "neo-Kraepelinian" nature of the *DSM-III*. E.g., Klerman, 1978; Compton and Guze, 1995; Schnittker, 2017, 207.
4. Spitzer, 1982.
5. Hempel, 1965. Physicist P. W. Bridgman first developed this model in 1927.
6. APA, 1980, 6.
7. Wakefield, 2001.
8. APA, 1980, 228.
9. Shorter, 2013.
10. Herzberg, 2009, 9.
11. The labels of "psychogenic" in the *DSM-I* and of "not attributed to physical conditions listed previously" in the *DSM-II*, however, did not convey the possibly biological nature of melancholic depression.
12. See especially Shorter, 2013.
13. Akiskal et al., 1978, 759.
14. Herzberg, 2009, 153.
15. Kadushin, 1969, 3.
16. Herzberg, 2009, 157.
17. Herzberg, 2009, 260n26. The revelation in 1972 that Democratic vice presidential nominee Thomas Eagleton was hospitalized and given shock treatments in the 1960s was the first event that put depression in the public spotlight.
18. Herzberg, 2009, 155, 166.
19. Maxmen, 1985, 46.
20. APA, 1968, 36.
21. Klein quoted in Shorter, 2013, 136.
22. Spitzer quoted in Shorter, 2013, 136. The *DSM-III* (1980, 205) contains a footnote to the affective disorders category: "Since the term 'endogenous' implies, to many, the absence of precipitating stress, a characteristic not always associated with this syndrome, the term 'endogenous' is not used in DSM-III."
23. A psychiatrist, Alan J. Eisnetz, had written a long letter to the president of the APA New York County chapter saying it was a "serious mistake to blur the distinction between the neurotic depressions and psychotic episodes," but his plea went unheeded. Eisnetz to Campbell, February 9, 1976, Archives, DSM Coll.
24. APA, 1980, 213–14. Spitzer initially proposed a necessary duration of just one week. Spitzer, Endicott, & Robins, 1975, 1189.
25. Lehmann, 1959, S3.
26. APA, 1980, 213, 333.
27. Horwitz & Wakefield, 2007, 100–103.
28. APA, 1980, 215. E.g., Roth, 1990.
29. Shorter, 2009, 160; Shorter, 2013, 137.
30. Spitzer indicated that this diagnosis stemmed from the results of the field trials where participating psychiatrists distinguished chronic depressive disorder from major

depressive disorder. He therefore argued for a new category of dysthymic disorder that is of lesser intensity than MDD (Spitzer to DSM-III Task Force, April 30, 1979, Archives, DSM Coll.). Yet he did not note that this category contained only those less severe depressions that are also very prolonged. Brief conditions would still quality for an MDD diagnosis.

31. Akiskal, 2001.

32. Shorter, 2013, 80.

33. Kessler et al., 2005.

34. D. Zamora, "Mental disorders common in America," WebMD, no date, https://www.webmd.com/mental-health/features/mental-disorders-common-in-america#2.

35. Kramer, 2005, 215.

36. Murray & Lopez, 1996; Institute of Medicine, 2001.

37. USDHHS, 1999.

38. Auden, 1947/1994.

39. Lane, 2007, 75–76.

40. APA, 1980, 225–39. Shorter (2013, 52) contends the sharp lines the manual drew around these supposedly independent conditions were "like trying to draw lines in a bucket of water."

41. APA, 1980, 232–33. The manual specifically mentions phobias, panic attacks, obsessions and compulsions, or MDD as examples.

42. Tyrer, 1984, 79.

43. Horwitz, 2013, 76–81; see also Lane 2007, 32–33. Spitzer was aware of this aspect of Freud's work and proposed that the introduction to the *DSM-III* mention that Freud used the term "psychoneuroses" descriptively as well as etiologically (Spitzer to American Psychoanalytic Association Liaison Committee, March 27, 1979, Archives, DSM Coll.).

44. Freud, 1894/1959, 79.

45. Shorter, 2013, 110.

46. World Health Organization, 1979, code 300.4.

47. Laing, 1967; Rosenhan, 1973.

48. Jean Endicott to Robert Spitzer, no date, Archives, DSM Coll.

49. Cooper et al., 1972.

50. Robins & Helzer, 1986, 411.

51. Endicott & Spitzer, 1978.

52. APA, 1968, 6–7.

53. APA, 1968, 33.

54. When a schizophrenic-seeming state persisted for less than six months, it received a diagnosis of schizophreniform disorder (APA, 1980, 187).

55. Skodol, 2000, 440.

56. Klein quoted in Shorter, 2009, 164.

57. The task force spent more time debating this issue than perhaps any other save for the controversy over neurosis

58. Cancro, 2000, 419.

59. APA, 1980, 202.

60. Ten percent of diagnoses were of pathological personality, 10 percent of some other character disorder, and 4 percent of immature personality (*NDTI Review*, 1970, 2, 1–5, 3).

61. Beck quoted in Decker, 2013, 196.

62. Richard A. Schwartz to Spitzer, no date, Archives, DSM Coll.

63. G. Schiff to Spitzer, January 28, 1979, Archives, DSM Coll.

64. Specific developmental disorders were also placed on Axis II.

65. APA, 1980, 305.

66. APA, 1980, 309, 311, 315, 318, 326, 329.

67. APA, 1980, 321.

68. Ted Millon to Spitzer, June 28, 1978, Archives, DSM Coll.

69. Donald Klein to Spitzer, November 9, 1978, Archives, DSM Coll.

70. Quoted in Decker, 2013, 199.

71. APA, 1980, 314–15.

72. APA, 1980, 324–25.

73. William E. Holt & Charles K. Hofling to Spitzer, December 12, 1975, Archives, DSM Coll. Spitzer was evidently not impressed by the analysts' arguments, calling one an "amiable anachronism" and the other a "well-known nobody."

74. Spitzer to Albert Mailwald, June 7, 1976, Archives, DSM Coll.

75. Judd Marmor to Spitzer, May 12, 1977, Archives, DSM Coll. In addition, some analysts pointed out the absurdity of limiting mental disorders to only distressing states: they contended that such a definition would eliminate from diagnosis such conditions as paranoia, schizophrenia, and manic depression.

76. Richard Green to Spitzer, December 14, 1976, Archives, DSM Coll.

77. APA, 1980, 282.

78. Bayer, 1987, 177.

79. APA, 1968, 49.

80. See Horwitz, 2018, 80–106.

81. Shatan, 1972.

82. In 1972 the APA passed a resolution that asserted, "We find it morally repugnant for any government to exact such heavy costs in human suffering for the sake of abstract concepts of national pride or honor" (Jones & Wessely, 2005, 135).

83. APA, 1980, 236, 238.

84. See Horwitz, 2018, 102.

85. E.g., McHugh, 1999; Satel, 2010.

86. APA, 1968, 45–46.

87. APA, 1980, 163–179.

88. APA, 1968, 49–50.

89. APA, 1980, 35–99. This does not count conditions in the mental retardation or developmental disorder categories.

90. APA, 1980, 89–92.

91. Skodol, 2000, 439.

92. Schnittker, 2017, 182–83, 190–93, 201–3.

93. Kendell, 1983, 66.

94. Klerman, 1984, 542. One psychologist quoted in Caplan (1995, 282) reputedly wrote that his colleagues "fell on their knees and begged to be included for reimbursement."

95. Kirk & Kutchins, 1992, 11.

96. Miller, Bergstrom, Cross, & Grube, 1981.

97. Paul J. Fink to Lester Grinspoon, May 15, 1978, Archives, DSM Coll.

98. Hale, 1995, 329.

99. Smith, 2019, 167. Psychiatrist Jeffrey Lieberman's obsession with psychoanalysis in *Shrinks* (2015) is an exception.

100. Maxmen, 1985, 14, 31, 32.

101. Prominent behavioral psychologist Hans Eysenck held a contrary view: "It should be recognized that DSM-III is primarily a *professional* manual and only secondarily, if at all, a *scientific* manual." He also observed that "psychologists may have to use the system because of social pressures of various kinds, but this should not blind them to the fundamental weaknesses of any such scheme based on democratic voting procedures rather than on scientific evidence" (Eysenck, Wakefield, & Friedman, 1983, 184, 189).

CHAPTER FIVE: **The *DSM-III-R* and *DSM-IV***

1. Schwartz & Wiggins, 2002.

2. First quoted in Greenberg, 2013, 67. Greenberg (2010) also quotes a former APA president who responds to his question about the value of a diagnosis, "I got paid."

3. Dobransky, 2014, 60.

4. E.g., Faust & Miner, 1986; Eysenck, Wakefield, & Friedman, 1983.

5. Kendell, 1991; Sadler, 2005, 156; Kirk & Kutchins, 1992, 209.

6. Spitzer quoted in Greenberg, 2013, 41.

7. Greenberg, 2010.

8. This book does not discuss the *DSM-IV-TR*.

9. Frances, 2013, 172. The fact that Spitzer maintained control over the revision might explain why it was called *DSM-III-R* instead of *DSM-IV*.

10. Some critics objected that the revised manual's many alterations in such a short period created a moving target for researchers and a new system for clinicians to master. The head of the *DSM-IV* revision, Allen Frances (2013, 68), complained that "DSM-III-R was a mistake and a distraction."

11. APA, 1987, 8. Blashfield et al. state, "The DSM-III-R was not just a revision: it was a new classification system" (2014, 34). Similarly, Kirk and Kutchins observe that "DSM-III-R constituted a major, not a minor change" (1992, 139). Yet, although the revision added some categories and made changes in others, in every essential aspect the two manuals seem comparable.

12. Frances (2013, 70) marveled at the casual and superficial way he was simply asked whether he would accept appointment to this weighty position. Similarly, after he assumed his role, the APA provided no direction for how he should proceed or whom he should choose for the various work groups.

13. Nathan, 1994, 105.

14. Frances, 2013, 69, 71–72.

15. See Sadler, 2005, 382–84, for a detailed description of the *DSM-IV* revision process.

16. Wakefield, 1996.

17. Spitzer, 1991.

18. Cotton, 1993.

19. APA, 1994, xxv.

20. Frances et al., 1991, 408.

21. Schnittker, 2017, 180–87.

22. Simultaneously with the development of the *DSM-III*, the NIMH and researchers affiliated with the Washington University group launched the first study that measured the prevalence of specific mental disorders in the community (Robins & Regier, 1991).

23. Kessler & Wang, 2008.

24. Grant et al., 2004.

25. Persons with serious mental illnesses met neither standard and accordingly became less visible.

26. Horwitz, 2002, 75–78.

27. Spitzer et al., 1989; Figert, 1996.

28. American Psychological Association Committee on Women quoted in Figert, 1996, 41. Another noted feminist writer, Carol Travis (1993, B7), later asserted, "In bestowing its self-serving approval to this label, the psychiatric Establishment feeds the prejudice that women's hormones, but not men's, are a cause of mental illness."

29. Figert, 1996, 44–45.

30. Notes of Ad Hoc Advisory Committee, June 3, 1985, Archives, DSM Coll.

31. Spitzer in the *Psychiatric Times*, August 1986, quoted in Figert, 1996, 93.

32. Spitzer to Advisory Committee on Premenstrual Dysphoric Disorder, October 28, 1985, Archives, DSM Coll.

33. Caplan, 1995, 8, 87–89.

34. Caplan, 1995, 89.

35. Caplan, 1995, 93.

36. Notes of Ad Hoc Advisory Committee, June 3, 1985, Archives, DSM Coll.

37. APA, 1987, 370–71.

38. Blumenthal & Nadelson, 1988.

39. The *Washington Times* (July 4, 1986) headlined its discussion of the resolution "APA Cuts Manual to Please Feminists" (Figert, 1996, 60).

40. Figert, 1996, 146.

41. Zachar & Kendler, 2014.

42. Figert, 1996, 162; Caplan, 1995, 270–71. Caplan (109) asserts, "When the DSM-III-R was published, the fact that the task force had baldly lied became clear. Although [it was] included in the provisional appendix, LLPDD was also listed in the main text of the manual, which is supposedly reserved for fully tested and scientifically supported diagnoses." In fact, this "bald lie" was the *DSM* statement under the category of unspecified mental disorder (nonpsychotic): "This category can also be used for specific mental disorders that are not included in the DSM-III-R classification, for example, Late Luteal Phase Dysphoric Disorder" (1987, 363). This statement indicates, contrary to Caplan's assertion, that LLPDD is "not included in the DSM-III-R classification."

43. APA, 1994, 715. The *DSM-IV* also confusingly lists PDD as an example of the category depressive disorder not otherwise specified (APA, 1994, 350).

44. Horwitz, 2018, ch. 5.

45. Dworkin, 1987, 194; MacKinnon, 1988; Steinem, 1983.

46. Bass & Davis, 1988. The remainder of this section is based on Grob & Horwitz, 2010, 179–82.

47. Herman, 1995, 1.

48. See especially McNally, 2003; Horwitz, 2018, ch. 5.

49. Borch-Jacobsen, 1997.

50. APA, 1980, 253, 259, 258.

51. APA, 1987, 271.

52. Acocella, 1999, 66, 68.

53. Hansen, 1998; Ofshe & Watters, 1994.

54. Acocella, 1999, 103.

55. This section is adopted from Horwitz, 2018, ch. 5 and ch. 6.

56. APA, 1952, 40.

57. APA, 1980, 238.

58. Sommers & Satel, 2006, 145.

59. Herman, 2015, 28.

60. APA, 1987, 250.

61. APA, 1994, 431–32; 427–28. In addition, one of the few new conditions in the *DSM-IV* was acute stress disorder (ASD), which allowed for a trauma-related diagnosis at any time within the first 30 days of the event. Because its symptoms needed to persist for only two days, ASD potentially pathologized a huge number of normal responses to stress.

62. Breslau & Kessler, 2001, 703.

63. Avina & O'Donohue, 2002, 74; McNally, Bryant, & Ehlers, 2003; Dattilio, 2004.

64. Breslau & Kessler, 2001, table 1, 701.

65. Galea, Nandi, & Vlahov, 2005, 83.

66. Brewin, Andrews, & Valentine, 2000, 752.

67. Norris et al., 2002, table 6, 230; table 2, 247.

68. Galea et al., 2007, table 2, 1430.

69. Horwitz & Wakefield, 2012, 176.

70. Boschen, 2008; Summerfield, 2001.

71. Ward, 2002, 207–10.

72. Lerner & Micale, 2001, 3.

73. Kramer, 1993, 85.

74. Shorter, 2013, 124.

75. Tone, 2009, 212–14. In 2013, more than 13 million Americans filled 49 million prescriptions for Xanax, the second highest of any psychoactive drug. E. MacLaren, "Xanax history and statistics," Drugabuse.com, no date, https://drugabuse.com/xanax/history-statistics/; Bachhuber et al., 2016.

76. Shorter, 2009, 149.

77. Herzberg, 2009, 165

78. The following six paragraphs are adapted from Horwitz, 2015.

79. See also Martino et al., 2019.

80. Klein to Spitzer, March 19, 1986, Archives, DSM Coll. See also Barlow, 1991.

81. Van Praag, 1990, 149.

82. Olfson et al., 2002a.

83. Olfson et al., 2002b; Mojtabai & Olfson, 2008.

84. CDC, National Center for Health Statistics, "Mental health," page last reviewed May 3, 2017, https://www.cdc.gov/nchs/fastats/mental-health.htm.

85. Schappert & Rechtsteiner, 2008.

86. Olfson et al., 2014.

87. Barber, 2008, 14.

88. Simons, 2004.

89. Herzberg, 2009, 177.

90. Olfson & Klerman, 1993.

91. Zuvekas, 2005.

92. Mojtabai & Olfson, 2008.

93. Raofi & Schappert, 2006.

94. Lane, 2007, 112.

95. Schnittker, 2017, 190–95.

96. See Shorter, 2009, 201.

97. Tone, 2009, 216–26.

98. The *DSM-IV* put a proposed category of mixed anxiety-depressive disorder in an appendix for conditions that warranted further study that the *DSM-5* subsequently eliminated.

99. See Lane, 2007, for an extensive discussion of social phobia.

100. The following paragraphs are adopted from Horwitz & Wakefield, 2012, 159–63.

101. APA, 1980, 228.

102. APA, 1980, 228.

103. Barlow, 1988, 536.

104. APA, 1987, 243.

105. APA, 1994, 411.

106. Magee et al., 1996.

107. Stein, Walker & Forde, 1994, 408.

108. Kessler & Wang, 2008, 123.

109. Horwitz & Wakefield, 2012, 161.

110. Lane, 2007, 105; Tone, 2009, 217.

111. Vedantam, 2001.

112. Moynihan & Cassels, 2005.

113. Lane, 2006, 388.

114. Frances, 2013, 153.

115. Winokur & Clayton, 1967.

116. APA, 1980, 217.

117. Frederick Goodwin and Kay Redfield Jamison's 1990 text, *Manic-Depressive Illness*, had argued against the *DSM-III*'s sharp split of bipolar from depressive conditions and for a return to Kraepelin's singular vision of them.

118. APA, 1994, 359, 338.

119. Wakefield, 1996, 650–51.

120. NIMH, "Bipolar disorder," last updated November 2017, https://www.nimh.nih.gov/health/statistics/bipolar-disorder.shtml; Bipolar Lives, "Bipolar disorder statistics," no date, https://www.bipolar-lives.com/bipolar-disorder-statistics.html.

121. "Battling bipolar," *People*, April 20, 2011; "9 must-see movies about bipolar disorder," *bp Magazine*, April 12, 2020, https://www.bphope.com/9-movies-about-bipolar-you-should-watch/.

122. Tone, 2009, 222.

123. APA, 1952, 24.

124. APA, 1968, 36.

125. The *DSM-III* dropped the involutional category.

126. Greenberg, 2013, 82; Harrington, 2019, 241.

127. Carey, 2007.

128. Moreno et al., 2007; Shorter, 2015, 85.

129. Groopman, 2007; Greenberg, 2013, 78–86; Frances, 2013, 144–46.

130. Harris & Carey, 2008.

131. Frances, 2013, 146.

132. Bayer, 1987, 210–11.

133. APA, 1987, 296. This category also included distress over repeated sexual conquests and feelings of inadequacy over one's sex organs.

134. Sadler, 2005, 206.

135. Frances et al., 1991, 411.

136. Kendell & Jablensky, 2003, 4.

CHAPTER SIX: **The *DSM-5*'s Failed Revolution**

1. The APA changed the numbering of the manual from roman to arabic in anticipation of numerous future revisions of the *DSM*. Henceforth, each revised version was to be 5.1, 5.2, etc. The APA wanted to brand the DSM as embodying a "continuous improvement model" that constantly progressed as new scientific information emerged (Zachar, First, & Kendler, 2017, 903). According to David Kupfer, the renumbering would lead to "a dynamic and living document and we don't have to wait 15 or 20 years, like with the previous DSM editions, before a mistake can be repaired" (Verhoef, 2010, 471).

2. Lieberman, 2015, 278.

3. Kupfer, First, & Regier, 2002, xix.

4. Zachar, Regier, & Kendler, 2019, 779.

5. Sharfstein, 2005.

6. Society for Neuroscience, "Chapter V: Neuroscience ascendant, mid-1980s to 1995," no date, https://www.sfn.org/about/history-of-sfn/1969-2019/chapter-5. By 2018 the Society for Neuroscience had more than 40,000 members.

7. Jasanoff, 2018, 20.

8. Andreasen, 1984, 30.

9. This section is adapted from Horwitz, 2020, 228–34.

10. Hyman, 2008, 890; Hyman, 2018, 944.

11. Wong, Choy, & Van Tol, 2003, 292.

12. Sprooten et al., 2017. See also Pettersson, Larsson, & Lichtenstein, 2016.

13. Gershon & Nurnberger, 1995; Weissman et al., 2006.

14. Van Dam, Iacoviello, & Murrough, 2018, 290.

15. LeDoux, 1998, 230.

16. McGlashan et al., 2000.

17. Krueger, 1999.

18. At the end of his career, even Kraepelin had started to question the sharp earlier division he had made between these two entities.

19. Duan, Sanders, & Gejman, 2010; Cross-Disorder Group of the Psychiatric Genome Consortium, 2013.

20. Fanous & Kendler, 2008.

21. Caspi & Moffitt, 2018, 637.

22. Nestler, 2018, 382.

23. Davidson, 2013, 206–7.

24. Robins & Guze, 1970.

25. Steve Mirin quoted in Greenberg, 2013, 60.

26. Ledford, 2009, 445.

27. Regier et al., 2009.

28. Regier et al., 2009, 646.

29. Regier et al., 2011, xvii.

30. First quoted in Greenberg, 2013, 74.

31. See especially Verhoeff, 2010.

32. APA, "*Diagnostic and statistical manual of mental disorders (DSM-5),*" http://www.dsm5.org/about/pages/dsmvoverview.aspx.

33. Ledford, 2009, 445.

34. Cosgrove et al., 2006.

35. Waldon, 2014.

36. Lieberman, 2015, 275.

37. Schatzberg et al., 2009.

38. First quoted in Greenberg, 2013, 92.

39. *Carlat Psychiatry Blog*, "Psychiatry's *DSM-V* process: Now a bar room brawl," June 30, 2009, http://carlatpsychiatry.blogspot.com/2009/06/psychiatrys-dsm-v-process-now-bar-room.html.

40. About 15 years after the *DSM-III* field trials, Kirk and Kutchins (1992) wrote a biting critique of them.

41. Zachar, Regier, & Kendler, 2019. Although the trials intended to enroll 5,000 clinicians, only a few more than 600 participated (Mościcki et al., 2013).

42. Kappas worse than chance agreement could reach a value of −1.

43. Jones, 2012, 518.

44. Regier et al., 2013.

45. A recent survey of meta-analyses also indicates that replication rates of psychiatric diagnoses across studies are appallingly low. See Dumas-Mallet et al., 2016.

46. Kraemer quoted in Greenberg, 2013, 312.

47. Kraemer et al., 2012a.

48. Narrow et al., 2013, 78.

49. Freedman et al., 2013.

50. Spitzer, Williams & Endicott, 2012.
51. Kraemer et al., 2012b.
52. Zachar, Regier, & Kendler, 2019.
53. Zachar, Regier, & Kendler, 2019, 780.
54. Whooley, 2016, 36.
55. For example, Jane Costello, a member of the Child and Adolescent Disorders work group, resigned because she was "increasingly uncomfortable with the whole underlying principle of rewriting the entire psychiatric taxonomy at one time. I am not aware of any other branch of medicine that does anything like this" (Whooley & Horwitz, 2013, 84).
56. Carolyn Robinowitz quoted in Lieberman, 2015, 279.
57. Zachar, Regier, & Kendler, 2019, 781, 782. In May 2012, yet another layer was added to the structure when APA president-elect Jeffrey Lieberman established a Summit Group to reevaluate the proposals that emerged from the various other review groups.
58. Mitchell quoted in Scull, 2019, 190.
59. Phalen et al., 2018.
60. The *DSM-III* had mentioned a prodromal phase of schizophrenia that involved "a clear deterioration in functioning before the active phase of the illness," but it applied to people who would have already met the full criteria for schizophrenia at a later time. It could be determined only after—not before—someone became psychotic.
61. Fusar-Poli et al., 2010; Verhoeff, 2010, 470.
62. Kessler et al., 2003.
63. Corcoran, First, & Cornblatt, 2010.
64. Carpenter, 2009, 842.
65. Weiden, 2012.
66. Frances, 2013, viii.
67. E.g., Kupfer in Verhoeff, 2010, 470; Fusar-Poli et al., 2010.
68. APA, 2013, 783–84.
69. Frances, 2013, 198.
70. Zachar, Regier, & Kendler, 2019.
71. Kendell & Jablensky, 2003. "For DSM-V, it's the 'disorder threshold,' stupid," Regier bluntly stated (Regier, Narrow, & Rae, 2004, 1051).
72. E.g., Costa & Widiger, 1993.
73. Kessler et al., 2003, 1118.
74. Regier et al., 2009, 648–49. The research base for dimensions, primarily stemming from studies that psychologists conducted, had the major limitation of being based on self-reported measures that couldn't be confirmed through any biomarker or other test. In addition, dimensional measurement would not help resolve the fundamental issue that Spitzer struggled with throughout his involvement in the DSM process: how to distinguish normal-range responses from mental disorders. Nor would a move to a dimensional system overcome the problems of the extant DSM system that neuroscientific and genetic research had identified. Treating diagnosis as continuous or categorical is irrelevant for importing genetic etiologies into diagnostic systems, deciding how many dimensions or categories a classification should have, dealing with the overlap

between different diagnoses, resolving the internal heterogeneity of many *DSM* conditions, determining how broadly or narrowly intervals on a dimension should be measured, or identifying what constitutes a disordered trait in the first place.

75. Zachar & Krueger, 2013.

76. Zachar & Krueger, 2013, 899.

77. Mischel, 1968, 1.

78. First, 2005; Trull & Widiger, 2013.

79. Skodol, 2012.

80. Phillips, 2011.

81. Kendler et al., 2009.

82. Pilkonis et al., 2011, 4.

83. Phillips, 2011.

84. The work group reversed its initial recommendation to delete narcissistic personality disorder.

85. Zachar, Kreuger, & Kendler, 2016, 4.

86. APA, 2013, 761–81.

87. Zachar, Krueger, & Kendler, 2016, 1.

88. Millon, 2012, 2.

89. Whooley, 2016, 37.

90. APA, 2013, 73; Whooley & Horwitz, 2013.

91. T. Insel, "Transforming diagnosis," NIMH, April 29, 2013, https://www.nimh .nih.gov/about/directors/thomas-insel/blog/2013/transforming-diagnosis.shtml.

92. Insel quoted in Greenberg, 2013, 340.

93. Insel, "Transforming diagnosis."

94. Horwitz & Wakefield, 2007.

95. Parker et al., 2010; See also Parker, McClure, & Paterson, 2015.

96. Greenberg, 2013, 337.

97. Greenberg, 2013, 337.

98. APA, 1987, xxii; APA, 1994, xxi.

99. Wakefield et al., 2007; Horwitz & Wakefield, 2007; Mojtabai, 2011; Wakefield & Schmitz, 2012; Wakefield & Schmitz, 2013a. Wakefield & Schmitz, 2013b.

100. Wakefield & First, 2012.

101. APA, *DSM-5* website, https://www.psychiatry.org/psychiatrists/practice/dsm.

102. APA, 2013, 161.

103. Carey, 2012b.

104. Kendler et al., 2009, 5.

105. E.g., Fawcett, 2010.

106. Haslam, 2016, 12–13.

107. APA, 1952, 15, 16, 20, 39; APA, 1968, 24, 31, 45–46.

108. APA, 1994, 181.

109. Wakefield, 2015a.

110. APA, 1994, 182. Because substance abuse could not be diagnosed when substance dependence was also present, it was a residual diagnosis.

111. My analysis is indebted to Wakefield, 2015b.

112. APA, 2013, 483.

113. Hasin, 2012, 703. At the same time, Hasin indicated that social pressure might have been a more important source than "overwhelming evidence." She stated that she changed her opposition to the conflation of dependence and abuse; after looking "around the table at my colleagues in our in-person meeting," she "no longer had grounds to hold to my position."

114. APA, 2013, 490–91.

115. Hasin et al., 2013; Wakefield, 2015b, 3.

116. Mewton et al., 2010.

117. APA, 2013, 571.

118. In addition, the *DSM-IV* had noted 15 types of over-the-counter and prescribed medications that could lead to the development of a substance-related disorder. The *DSM-5* SUD diagnosis, however, excluded medically prescribed substances from diagnosis: "Symptoms of tolerance and withdrawal occurring during appropriate medical treatment with prescribed medications (e.g., opioid analgesics, sedatives, stimulants) are specifically *not* counted when diagnosing a substance use disorder" (APA, 2013, 484). Identical symptoms that indicated disorders when they resulted from taking illicit drugs were not disorders when they stemmed from a prescribed drug. This change was spectacularly untimely, implemented just as medically prescribed opioids were producing perhaps the worst epidemic of addiction in American history. Yet if clinicians adhered to the new criteria, they could not use the SUD diagnosis to provide treatment to addicted opioid users.

119. Allen Frances also noted how the definition lumps first-time abusers with hard-core addicts despite their very different etiologies, prognoses, and treatment needs (Mechanic, 2013, 7).

120. Edwards, 2012, 701.

121. APA, 2013, 50–52.

122. Lord & Bishop, 2015.

123. E.g., Charman, 2011.

124. McPartland, Reichow, & Volkmar, 2012; Greenberg, 2013, 297.

125. Carey, 2012a.

126. Greenberg, 2013, 299.

127. Greenberg, 2013, 196.

128. APA, 2013, 51.

129. McPartland, Riechow, & Volkmar, 2012.

130. APA, 1980, 261–66; APA, 1987, 71–77; APA, 1994, 532–38.

131. Greenberg, 2013, 101.

132. Greenberg, 2013, 100–101.

133. Greenberg, 2013, 101.

134. APA, 2013, 452–53.

135. Lieberman, 2012.

136. Judith Rapoport quoted in Grinker, 2007, 131.

137. First, 2005, 562.

138. Frances quoted in Mechanic, 2013, 5.

CHAPTER SEVEN: **The *DSM* as a Social Creation**

1. National Association of State Mental Health Program Directors, 2017.
2. Cf. Meehl, 1972, 64.
3. Mirowsky & Ross, 2002, 160.
4. Andreasen, 1995, 964.
5. Scull, 2019, 214.
6. Scull, 2019, 290.
7. Cosgrove & Krimsky, 2012.
8. Ferdman, 2013; Moreno et al., 2007; Olfson et al., 2006.
9. Conrad, 2005; Haslam, 2016.
10. Frances, 2013, xii.
11. Grinker, 2007, 171.
12. Spitzer quoted in Kirk & Kutchins, 1992, 192.
13. Grob, 1987, 417.
14. APA, 1952, 28, 41–42; APA, 1968, 49–51.
15. APA, 1980, 35–100.
16. CDC, "Attention deficit / hyperactivity disorder," page last reviewed October 15, 2019, https://www.cdc.gov/ncbddd/adhd/data.html.
17. Lord & Bishop, 2015. A simple editorial error in the *DSM-IV* contributed to the explosion of autism diagnoses after 1994. The work group intended to write "impairment in social interaction *and* in verbal or nonverbal communication skills," but the manual published the criteria as "impairment of reciprocal social interaction *or* verbal and nonverbal communication skills, *or* when stereotyped behavior, interests, and activities are present." This error persisted for six years until the *DSM-IV-TR* corrected it in 2000 (Grinker, 2007, 140–41).
18. Angst, 1998.
19. J. Condliffe, "The average American spends 24 hours a week online," *MIT Technology Review*, January 23, 2018, https://www.technologyreview.com/f/610045/the-average-american-spends-24-hours-a-week-online/; M. Anderson et al., "10% of Americans don't use the internet: Who are they?" Pew Research Center, April 22, 2019, https://www.pewresearch.org/fact-tank/2019/04/22/some-americans-dont-use-the-internet-who-are-they/; S. Salim, "How much time do you spend on social media? Research says 142 minutes per day," Digital Information World, January 4, 2019, https://www.digitalinformationworld.com/2019/01/how-much-time-do-people-spend-social-media-infographic.html.
20. Anderson, 2018.
21. Haslam, 2016, 8.
22. Menninger, 1963, 33.
23. APA, 1968, 39.
24. Wakefield, 2001, 590.
25. The *DSM-I* (1952, 34–39) split the 12 personality disorders into four personality pattern disturbances, four personality trait disturbances, and four sociopathic personality disturbances.

26. APA, 2013, 645.

27. Haslam, 2016, 12.

28. APA, 2013, 271–72.

29. Harris & Carey, 2008; Harris, 2009; Couric, 2007.

30. APA, 2013, 156.

31. Critics, however, were still not convinced that this relabeling resolved the problem of overmedication. Allen Frances (2012b), for example, remained skeptical: "We have no idea whatever how this untested new diagnosis will play out in real life practice settings."

32. E.g., Frances, 2013; Rose, 2006; Williams, Martin & Gabe, 2011.

33. Grob, 1998, 217.

34. Kendler, 1990, 969.

35. APA, "About *DSM-5*," https://www.psychiatry.org/psychiatrists/practice/dsm /about-dsm.

36. Whooley, 2019.

37. Shorter, 2009, 4. Psychiatrist Simon Wessely (2005, 549) pushes this time frame back even further for PTSD: "There is probably little we could now teach either the Regimental Medical Officers of the First World War, or the psychiatrists of the Second, about the psychological effects of war."

38. Arnold quoted in Lewis, 1967, 193.

39. Kendell & Jablensky, 2003, 9.

40. Pies, 2008, 12.

41. Noll, 2011, 175.

42. Kendler, 2008.

43. Kleinman, 1988.

44. Paris, 2015. A recent example is the Centers for Disease Control's 2020 recommendation that everyone should wash their hands numerous times a day in response to the epidemic of COVID-19. In other contexts, this behavior could seem like the ritualistic handwashing that characterizes obsessive-compulsive disorder.

45. APA, 2013, 20.

46. The harmful dysfunction definition of mental disorder has drawn much criticism. Some argue that dysfunctions in themselves are sufficient indicators of disorder, so any notion that they must also be harmful is unnecessary (e.g., Klein, 1999). Other critics make the opposite objection, contending that distressing or impairing psychological conditions should be treated, regardless of their dysfunctional qualities (e.g., Bolton, 2008). A third position holds that any general definition of mental disorder is hopeless. For example, Allen Frances scorns efforts to define mental disorder and calls the *DSM* definition "bullshit, I mean you can't define it" (quoted in Greenberg, 2013, 23). Yet his critique of the *DSM* rests on notions such as "out-of-control psychiatric diagnosis," "protecting normality from medicalization," or resisting "false epidemics of mental disorder" that necessarily assume some notion of what a genuine disorder is. Similarly, the British Psychological Society launched a highly publicized attack on the *DSM*, charging it with engaging in "the continued and continuous medicalisation of . . . natural and normal responses." ("Society's critical response to DSM-5," *the psychologist* [blog], https://thepsychologist.bps.org.uk/volume-24/edition-8/news). Yet such accusations require some definition of "natural" and "normal" that the BPS does not provide.

Despite these objections, the *DSM*'s definition of "mental disorder" provides a useful guide for developing specific diagnostic criteria sets.

47. For example, the first sentence of the anxiety disorders chapter states, "Anxiety disorders include disorders that share features of excessive fear and anxiety and related behavioral disturbances" (APA, 2013, 189). All eight of its specific criteria sets contain terms such as "excessive," "out of proportion," or "unexpected" that indicate they are not natural responses to dangerous situations. Such qualifiers are inherently opaque yet still necessary.

48. Wakefield & First, 2012.

49. APA, 1952, 34; APA, 1968, 40.

50. APA, 2013, 160–61. As the previous chapter indicated, while the MDD criteria set does not mention context, a note and footnote suggest that clinicians consider whether symptoms arise as responses to a significant loss.

51. APA, 2013, 248.

52. Cooper, 2014.

53. APA, 2013, 715.

54. APA, 2015, 749, 224, 103.

55. Earlier, psychiatrist Peter McHugh (1999) grouped symptoms into four classes of brain damage, uncontrollable addictions, and cognitive and emotional impairments.

56. Bracken et al., 2012, 430.

References

Acocella, J. (1999). *Creating hysteria: Women and multiple personality disorder.* San Francisco: Jossey-Bass.

Akiskal, H. S. (2001). "Dysthymia and cyclothymia in psychiatric practice a century after Kraepelin." *Journal of Affective Disorders, 62,* 17–31.

Akiskal, H. S., Bitar, A. H., Puzantian, V. R., Rosenthal, T. L., & Walker, P. W. (1978). "The nosological status of neurotic depression." *Archives of General Psychiatry, 35,* 756–66.

American Psychiatric Association (APA). (1952). *Diagnostic and statistical manual of mental disorders.* Washington, DC: Author.

American Psychiatric Association (APA). (1968). *Diagnostic and statistical manual of mental disorders* (2nd ed.). Washington, DC: Author.

American Psychiatric Association (APA). (1974). *Diagnostic and statistical manual of mental disorders* (2nd ed., 7th printing). Washington, DC: Author.

American Psychiatric Association (APA). (1980). *Diagnostic and statistical manual of mental disorders* (3rd ed). Washington, DC: Author.

American Psychiatric Association (APA). (1987). *Diagnostic and statistical manual of mental disorders* (3rd ed. rev.). Washington, DC: Author.

American Psychiatric Association (APA). (1994). *Diagnostic and statistical manual of mental disorders* (4th ed.). Washington, DC: Author.

American Psychiatric Association (APA). (2000). *Diagnostic and statistical manual of mental disorders* (4th ed., text rev.). Washington, DC: Author.

American Psychiatric Association (APA). (2010). Rationale for changes to substance-related disorders—tentative new title of new combined section: Addiction and related disorders. www.dsm5.org/ProposedRevision/Pages/proposedrevision.aspx?rid=431#.

American Psychiatric Association (APA). (2013). *Diagnostic and statistical manual of mental disorders* (5th ed.). Washington, DC: Author.

Anderson, J. (2018). "Even teens are worried they spend too much time on their phones." *Quartz,* August 23. https://qz.com/1367506/pew-research-teens-worried-they-spend-too-much-time-on-phones/.

Andreasen, N. (1984). *The broken brain: The biological revolution in psychiatry.* New York: Harper & Row.

Andreasen, N. (1995). "Posttraumatic stress disorder: Psychology, biology, and the Manichean warfare between false dichotomies." *American Journal of Psychiatry, 152,* 964.

Andrews, G., Hobbs, M. J., Borkovec, T. D., Beesdo, K., Craske, M. G., & Heimberg, R. G. (2010). "Generalized worry disorder: A review of DSM-IV generalized anxiety disorder and options for DSM-V." *Depression and Anxiety, 27,* 134–47.

Angell, M. (2009). "Drug companies & doctors: A story of corruption." *New York Review of Books,* January 15, 29–33.

Angst, J. (1998). "The emerging epidemiology of hypomania and bipolar II disorder." *Journal of Affective Disorders, 50,* 163–73.

Archive of the American Psychiatric Association Foundation. Washington, DC.

Auden, W. H. (1947/1994). *The age of anxiety.* Cutchoque, NY: Buccaneer Books.

Avina, C., & O'Donohue, W. (2002). "Sexual harassment and PTSD: Is sexual harassment diagnosable trauma?" *Journal of Traumatic Stress, 15,* 69–75.

Bachhuber, M. A., Hennessy, S., Cunningham, C. O., & Starrels, J. L. (2016). "Increasing benzodiazepine prescriptions and overdose mortality in the United States, 1996–2013." *American Journal of Public Health, 106*(4), 686–88. https://doi.org/10.2105/AJPH.2016.303061.

Barber, C. (2008). *Comfortably numb: How psychiatry is medicating a nation.* New York: Pantheon.

Barlow, D. H. (1988). *Anxiety and its disorders.* New York: Guilford Press.

Barlow, D. H. (1991). "Disorders of emotions: Clarification, elaboration, and future directions." *Psychological Inquiry, 2,* 97–105.

Bass, E., & Davis, L. (1988). *The courage to heal: A guide for women survivors of child sexual abuse.* New York: Harper and Row.

Batelaan, N. M., Spijker, J., de Graaf, R., & Cuijpers, P. (2012). "Mixed anxiety depression should not be included in DSM-5." *Journal of Nervous and Mental Disorders, 200,* 495–98.

Bayer, R. (1987). *Homosexuality and American psychiatry.* Princeton, NJ: Princeton University Press.

Bayer, R., & Spitzer, R. L. (1985). "Neurosis, psychodynamics, and DSM-III: History of the controversy." *Archives of General Psychiatry, 42,* 187–96.

Beard, G. (1869). "Neurasthenia, or nervous exhaustion." *British Medical Society Journal, 80,* 217–21.

Beck, A. T., Ward, C. H., Mendelson, M., Mock, J. E., & Erbaugh, J. K. (1962). "Reliability of psychiatric diagnosis: 2. A study of consistency of clinical judgments and ratings." *American Journal of Psychiatry, 119,* 351–57.

Berkley, H. J. (1900). *A treatise on mental diseases.* New York: D. Appleton.

Blackwell, B. (1975). "Minor tranquilizers: Use, misuse, or overuse?" *Psychosomatics, 16,* 28–31.

Blashfield, R. K., Keeley, J. W., Flanagan, E. H., & Miles, S. R. (2014). "The cycle of classification: DSM-I through DSM-5." *Annual Review of Clinical Psychology, 10,* 25–51.

Bleuler, E. (1911). *Dementia praecox: Or the group of schizophrenias.* New York: International Universities Press.

Blumenthal, S., & Nadelson, C. (1988). "Late luteal phase dysphoric disorder (premenstrual syndromes): Clinical implications." *Journal of Clinical Psychiatry, 49,* 469–74.

Bolton, D. (2008). *What is mental disorder?* New York: Oxford University Press.

Borch-Jacobsen, M. (1997). "Sybil: The making of a disease? An interview with Dr. Herbert Spiegel." *New York Review of Books*, April 24.

Boschen, M. J. (2008). "Publication trends in individual anxiety disorders: 1980–2015." *Journal of Anxiety Disorders, 22*, 570–75.

Bracken, P., Thomas, P., Timimi, S., Asen, E., Behr, G., Beuster, C., et al. (2012). "Psychiatry beyond the current paradigm." *British Journal of Psychiatry, 201*, 430–34.

Breslau, N., & Kessler, R. C. (2001). "The stressor criterion in DSM-IV posttraumatic stress disorder: An empirical investigation." *Biological Psychiatry, 50*, 699–704.

Brewin, C. R., Andrews, B., & Valentine, J. D. (2000). "Meta-analysis of risk factors for posttraumatic stress disorder in trauma-exposed adults." *Journal of Consulting and Clinical Psychology, 68*, 748–66.

Burnham, J. (2012). *After Freud left: A century of psychoanalysis in America.* Chicago: University of Chicago Press.

Cahalan, S. (2019). *The great pretender.* New York: Grand Central.

Callard, F. (2016). "The intimate geographies of panic disorder: Parsing anxiety through psychopharmacological dissection." *Osiris, 31*, 203–26.

Cancro, R. (2000). "Functional psychoses and the conceptualization of mental illness." In *American psychiatry after World War II: 1944–1994*, R. W. Menninger & J. C. Nemiah (eds.), 413–29. Washington, DC: American Psychiatric Press.

Caplan, P. J. (1995). *They say you're crazy: How the world's most powerful psychiatrists decide who's normal.* Reading, MA: Addison-Wesley.

Carey, B. (2007). "Debate over children and psychiatric drugs." *New York Times*, February 15. https://www.nytimes.com/2007/02/15/us/15bipolar.html.

Carey, B. (2012a). "New definition of autism will exclude many." *New York Times*, January 9, A1.

Carey, B. (2012b). "Grief could join list of disorders." *New York Times*, January 24, 2012. https://www.nytimes.com/2012/01/25/health/depressions-criteria-may-be-changed-to-include-grieving.html.

Carey, B., & Harris, G. (2008). "Psychiatric group faces scrutiny over drug industry ties." https://www.nytimes.com/2008/07/12/washington/12psych.html.

Carpenter, W. (2009). "Anticipating DSM-V: Should psychosis risk become a diagnostic class?" *Schizophrenia Bulletin, 35*, 841–43.

Caspi, A., & Moffitt, T. E. (2018). "All for one and one for all: Mental disorders in one dimension." *American Journal of Psychiatry, 175*, 831–44.

Charman, T. (2011). "The highs and lows of counting autism." *American Journal of Psychiatry, 168*, 873–75.

Clark, L. A. (ed.). (1999). "Special section: The concept of disorder: Evolutionary analysis and critique." *Journal of Abnormal Psychology, 108*, 371–472.

Compton, W. M., & Guze, S. B. (1995). "The neo-Kraepelinian revolution in psychiatric diagnosis." *European Archives of Psychiatry and Clinical Neuroscience, 245*, 196–201.

Conrad, P. (2005). "The shifting engines of medicalization." *Journal of Health and Social Behavior, 46*, 3–14.

Cooper, J., Rendell, R., Burland, B., Sharpe, L., Copeland, J., & Simon, R. (1972). *Psychiatric diagnosis in New York and London.* London: Oxford University Press.

Cooper, R. (2014). *Diagnosing the Diagnostic and Statistical Manual of Mental Disorders.* London: Karmac.

Cooper, R., & Blashfield, R. K. (2016). "Re-evaluating DSM-I." *Psychological Medicine,* 46, 449–56.

Corcoran, C. M., First, M. B., & Cornblatt B. (2010). "The psychosis risk syndrome and its proposed inclusion in the DSM-V: A risk-benefit analysis." *Schizophrenia Research,* 120, 16–22.

Cosgrove, L., & Krimsky, S. (2012). "A comparison of DSM-IV and DSM-5 panel members' financial associations with industry: A pernicious problem persists." *PLoS medicine,* 9(3), e1001190. https://doi.org/10.1371/journal.pmed.1001190. https://www.ncbi.nlm.nih.gov/pmc/articles/PMC3302834.

Cosgrove, L., Krimsky, S., Vijayaraghavan, M., & Schneider, L. (2006). "Financial ties between DSM-IV panel members and the pharmaceutical industry." *Psychotherapy and Psychosomatics,* 75, 154–60.

Costa, P. T., & Widiger, T. (eds.). (1993). *Personality disorders and the five-factor model of personality.* Washington, DC: American Psychological Association.

Cotton, P. (1993). "Psychiatrists set to approve DSM-IV." *Journal of the American Medical Association,* 270, 13–15.

Couric, K. (2007). "What Killed Rebecca Riley?" *60 Minutes,* September 30. https://www.cbsnews.com/video/what-killed-rebecca-riley/.

Cross-Disorder Group of the Psychiatric Genome Consortium. (2013). "Genetic relationship between five psychiatric disorders estimated from genome-wide SNPs." *Nature Genetics,* 45, 984–94.

Dain, N. (1964). *Concepts of insanity in the United States, 1789–1865.* New Brunswick, NJ: Rutgers University Press.

Dattilio, F. M. (2004). "Extramarital affairs." *Behavior Therapist,* 27, 76–78.

Davidson, L. (2013). "Cure and recovery." In *The Oxford handbook of philosophy and psychiatry,* K. W. M. Fulford et al. (eds.), 197–213. New York: Oxford University Press.

Davies, J. (2017). "How voting and consensus created the Diagnostic and Statistical Manual of Mental Disorders (DSM-III)." *Anthropology and Medicine,* 24, 32–46.

Decker, H. S. (2013). *The making of DSM-III.* New York: Oxford University Press.

Dobransky, K. (2014). *Managing madness in the community: The challenge of contemporary mental health care.* New Brunswick, NJ: Rutgers University Press.

Duan, J., Sanders, A. R., & Gejman, P. V. (2010). "Genome-wide approaches to schizophrenia." *Brain Research Bulletin,* 83, 93–102.

Dumas-Mallet, E., Button, K., Boraud, T., Munafo, M., & Gonon, F. (2016). "Replication validity of initial association studies: A comparison between psychiatry, neurology and four somatic diseases. *PLoS ONE,* 11, e0158064. doi:10.1371/journal.pone.0158064.

Dumit, J. (2004). *Picturing personhood: Brain scans and biomedical identity.* Princeton, NJ: Princeton University Press.

Dworkin, A. (1987). *Intercourse.* New York: Basic Books.

Edwards, G. (2012). "'The evil genius of the habit': *DSM-5* seen in historical context." *Journal of Studies on Alcohol and Drugs,* 73, 699–701.

Endicott, J., & Spitzer, R. L. (1978). "A diagnostic interview: The schedule for affective disorders and schizophrenia." *Archives of General Psychiatry, 35,* 837–44.

Everitt, B. S., Gourlay, A. J., & Kendell, R. E. (1971). "An attempt at validation of traditional psychiatric syndromes by cluster analysis." *British Journal of Psychiatry, 119,* 399–412.

Eysenck, H. J., Wakefield, J. A., & Friedman, A. F. (1983). "Diagnosis and clinical assessment: The *DSM-III.*" *Annual Review of Psychology, 34,* 167–93.

Fanous, A. H., & Kendler, K. S. (2008). "Genetics of clinical features and subtypes of schizophrenia." *Current Psychiatry Reports, 10,* 164–70.

Faust, D., & Miner, R. A. (1986). "The empiricist and his new clothes: DSM-III in perspective." *American Journal of Psychiatry, 143,* 962–67.

Fawcett, J. (2010). "Overview of mood disorders in the DSM-5. *Current Psychiatry Report, 12,* 531–38.

Feighner, J. P., Robins, E., Guze, S. B., Woodruff, R. A., Winokur, G., & Munoz, R. (1972). "Diagnostic criteria for use in psychiatric research." *Archives of General Psychiatry, 26,* 57–63.

Ferdman, Roberto A. (2013). "The alarming rise of Adderall in two charts." *Quartz,* December 17. https://qz.com/158457/the-alarming-rise-of-adderall-in-two-charts/.

Figert, A. E. (1996). *Women and the ownership of PMS: The structuring of a psychiatric disorder.* New York: Aldine de Gruyter.

Fine, R. (1979). *A history of psychoanalysis.* New York: Columbia University Press.

First, M. B. (2005). "Clinical utility: A prerequisite for the adoption of a dimensional approach in DSM." *Journal of Abnormal Psychology, 114,* 560–64.

Fleck, L. (1935/1979). *Genesis and development of a scientific fact.* Chicago: University of Chicago Press.

Foucault, M. (1965). *Madness and civilization: A history of insanity in the Age of Reason.* New York: Pantheon.

Frances, A. J. (2010). "Psychiatric diagnosis gone wild: The 'epidemic' of childhood bipolar disorder." https://www.psychologytoday.com/us/blog/dsm5-in-distress /201004/psychiatric-diagnosis-gone-wild-the-epidemic-childhood-bipolar-disorder.

Frances, A. J. (2012a). "Internet addiction—the next new fad diagnosis." https://www .psychiatrictimes.com/dsm-5/internet-addiction-next-new-fad-diagnosis

Frances, A. J. (2012b). "DSM-5 is a guide, not a bible—simply ignore its 10 worst changes." https://www.psychiatrictimes.com/alcohol-abuse/dsm-5-guide-not -biblesimply-ignore-its-10-worst-changes.

Frances, A. J. (2013). *Saving normal.* New York: William Morrow.

Frances, A. J., First, M. B., Widiger, T. A., Miele, G. M., Tilly, S. M., Davis, W. W., et al. (1991). "An A to Z guide to DSM-IV conundrums." *Journal of Abnormal Psychology, 100,* 407–12.

Freedman, R., Lewis, D. A., Michels, R., Pine, D. S., Schultz, S. K., Tamminga, C. A., et al. (2013). "The initial field trials of DSM-5: New blooms and old thorns." *American Journal of Psychiatry, 170,* 1–4.

Freud, S. (1894/1959). "The justification for detaching from neurasthenia a particular syndrome: The anxiety-neurosis." In *Collected papers,* vol. 1, Riviere, J. (trans.), 76–106. New York: Basic Books.

Friedan, B. (1963/2001). *The feminine mystique*. New York: W. W. Norton.

Fusar-Poli, P., Bonoldi, I., Yung, A. R., Borgwardt, S., Kempton, M. J., Valmaggia, L., et al. (2012). "Predicting psychosis: Meta-analysis of transition outcomes in individuals at high clinical risk." *Archives of General Psychiatry, 69,* 220–29.

Gabbard, G. O., & Gabbard, K. (1999). *Psychiatry and the cinema* (2nd ed.). Washington, DC: American Psychiatric Press.

Galea, S., Brewin, C. R., Gruber, M., Jones, R. T., King, D. W., King, L. A., McNally, R. J., Ursano, R. J., Petukhova, M., & Kessler, R. C. (2007). "Exposure to hurricane-related stressors and mental illness after Hurricane Katrina." *Archives of General Psychiatry, 64,* 1427–34.

Galea S., Nandi, A., & Vlahov D. (2005). "The epidemiology of post-traumatic stress disorder after disasters." *Epidemiologic Reviews, 27,* 78–91.

Galea, S., Vlahov, D., Resnick, H., Ahern, J., Susser, E., Gold, J., Bucuvalas, M., & Kilpatrick, D. (2003). "Trends of probable post-traumatic stress disorder in New York City after the September 11 terrorist attacks." *American Journal of Epidemiology, 158,* 514–24.

Gardner, E. 1971. "Psychoactive drug utilization." *Journal of Drug Issues, 1,* 295–300.

Gershon, E., & Nurnberger, J. I. (1995). "Bipolar illness." *Review of Psychiatry, 14,* 339–423. Washington, DC: American Psychiatric Press.

Goldberg, D., Kendler, K. S., Sirovatka, P. J., & Regier, D. A. (eds.) (2010). *Diagnostic issues in depression and generalized anxiety disorder*. Arlington: American Psychiatric Association.

Goodwin, F. K., & Jamison, K. R. (1990). *Manic-depressive illness*. New York: Oxford University Press.

Gorenstein, D. (2013). "How much is the *DSM-5* worth?" *Marketplace,* May 17. https://www.marketplace.org/2013/05/17/how-much-dsm-5-worth/.

Greenberg, G. (2010). "Inside the battle to define mental illness." *Wired,* December 27. http://www.wired.com/magazine/2010/12/ff_dsmv/all/1.

Greenberg, G. (2013). *The book of woe*. New York: Blue Rider Press.

Grant, B. F., Hasin, D. S., Stinson, F. S., Dawson, D. A., Chou, S. P., Ruan, W. J., & Pickering, R. P. (2004). "Prevalence, correlates, and disability of personality disorders in the United States: Results from the National Epidemiologic Survey on alcohol and related conditions." *Journal of Clinical Psychiatry, 65,* 948–58.

Grinker, R. R. (2007). *Unstrange minds: Remapping the world of autism*. New York: Basic Books.

Grob, G. N. (1987). "The forging of mental health policy in America: World War II to the new frontier." *Journal of the History of Medicine & Allied Sciences, 42,* 410–46.

Grob, G. N. (1990). "World War II and American psychiatry." *Psychohistory Review, 19,* 41–69.

Grob, G. N. (1991a). "Origins of DSM-I: Appearances and reality." *American Journal of Psychiatry, 148,* 421–31.

Grob, G. N. (1991b). *From asylum to community: Mental health policy in modern America*. Princeton, NJ: Princeton University Press.

Grob, G. N. (1994). *The mad among us: A history of the care of America's mentally ill*. New York: Free Press.

Grob, G. N. (1998). "Psychiatry's holy grail: The search for the mechanisms of mental disease." *Bulletin of the History of Medicine, 72,* 189–219.

Grob, G. N. (2011). "The attack of psychiatric legitimacy in the 1960s: Rhetoric and reality." *Journal of the History of the Behavioral Sciences, 47,* 398–416.

Grob, G. N., & Goldman, H. H. (2006). *The dilemma of federal mental health policy: Radical reform or incremental change?* New Brunswick, NJ: Rutgers University Press.

Grob, G. N., & Horwitz, A. V. (2010). *Diagnosis, therapy, and evidence: Conundrums in modern American medicine.* New Brunswick, NJ: Rutgers University Press.

Groopman, J. (2007). "What's the trouble: How doctors think." *New Yorker,* January 29.

Hackett, T. (1977). "The psychiatrist: In the mainstream or on the banks of medicine?" *American Journal of Psychiatry, 134,* 432–35.

Hale, N. (1995). *The rise and crisis of psychoanalysis in the United States.* New York: Oxford University Press.

Hamilton, M., & White, J. M. (1959). "Clinical syndromes in depressive states." *Journal of Mental Science, 105,* 985–98.

Hansen, C. (1998). "Dangerous therapy: The story of Patricia Burgus and multiple personality disorder." *Chicago Magazine,* June 1.

Harrington, A. (2019). *Mind fixers: Psychiatry's troubled search for the biology of mental illness.* New York: W. W. Norton.

Harris, G. (2009). "Drug maker told studies would aid it, papers say." *New York Times,* March 19. https://www.nytimes.com/2009/03/20/us/20psych.html.

Harris, G., & Carey, B. (2008). "Researchers fail to reveal full drug pay." *New York Times,* June 8. https://www.nytimes.com/2008/06/08/us/08conflict.html.

Hasin, D. S. (2012). "Combining abuse and dependence in DSM-5." *Journal of Studies on Alcohol and Drugs, 73,* 702–4.

Hasin, D. S., & Grant, B. F. (2015). "The National Epidemiologic Survey on Alcohol and Related Conditions (NESARC) waves 1 and 2: Review and summary of findings." *Social Psychiatry and Psychiatric Epidemiology, 50,* 1609–40.

Hasin, D. S., O'Brien, C. P., Auriacombe, M., Borges, G., Bucholz, K., Budney, A., et al. (2013). "DSM-5 criteria for substance use disorders: Recommendations and rationale." *American Journal of Psychiatry, 170,* 834–51.

Haslam, N. (2016). "Concept creep: Psychology's expanding concepts of harm and pathology." *Psychological Inquiry, 27,* 1–17.

Havermann, E. (1957). "The age of psychology in the U.S." *Life,* January 7, 68–72.

Healy, D. (1997). The anti-depressant era. Cambridge, MA: Harvard University Press.

Healy, D. (2000). *The pharmacologists III: Interviews.* New York: Oxford University Press.

Heckers, S. (2009). "Who is at risk for a psychotic disorder?" *Schizophrenia Bulletin, 35,* 847–50.

Hempel, C. G. (1965). "Fundamentals of taxonomy." In *Aspects of scientific explanation and other essays,* Hempel, C. G. (ed.), 137–54. Glencoe: Free Press.

Herman, E. (1995). *The romance of American psychology.* Berkeley: University of California Press.

Herman, J. (2015). *Trauma and recovery.* New York: Basic Books.

Herzberg, D. (2009). *Happy pills in America: From Miltown to Prozac.* Baltimore: Johns Hopkins University Press.

Herzberg, D. (2011). "Blockbusters and controlled substances: Miltown, Quaalude, and consumer demand for drugs in postwar America." *Studies in History and Philosophy of Science, 42,* 415–26.

Hill, C. G. (1907). "Presidential address." *American Journal of Insanity, 64,* 1–8.

Hinshaw, S. P., & Scheffler, R. M. (2014). *The ADHD explosion: Myths, medication, money, and today's push for performance.* New York: Oxford University Press.

Horwitz, A. V. (2002). *Creating mental illness.* Chicago: University of Chicago Press.

Horwitz, A. V. (2013). *Anxiety: A short history.* Baltimore: Johns Hopkins University Press.

Horwitz, A. V. (2015). "How did everyone get diagnosed with major depressive disorder?" *Perspectives in Biology and Medicine, 58,* 105–19.

Horwitz, A. V. (2018). *PTSD: A short history.* Baltimore: Johns Hopkins University Press.

Horwitz, A. V. (2020). *Between sanity and madness: Mental illness from ancient Greece to the neuroscientific era.* New York: Oxford University Press.

Horwitz, A. V., & Wakefield, J. C. (2007). *The loss of sadness.* New York: Oxford University Press.

Horwitz, A. V., & Wakefield, J. C. (2012). *All we have to fear: Psychiatry's transformation of natural anxiety into mental disorder.* New York: Oxford University Press.

Houts, A. C. (2000). "Fifty years of psychiatric nomenclature: Reflections on the 1943 War Department Technical Bulletin, Medical 203." *Journal of Clinical Psychology, 56,* 935–67.

Hyman, S. E. (2008). "A glimmer of light for neuropsychiatric disorders." *Nature, 455,* 890–93.

Hyman, S. E. (2018). "The present and future of psychiatric diagnosis." In *Charney & Nestler's neurobiology of mental illness* (5th ed.), Charney, D. S., Buxbaum, J. D., Sklar, P., & Nestler, E. J. (eds.), 941–46. New York: Oxford University Press.

Institute of Medicine Committee on Nervous System Disorders in Developing Countries (2001). *Neurological, psychiatric, and developmental disorders: Meeting the challenge in the developing world.* Washington, DC: National Academies Press.

Jackson, S. (1986). *Melancholia and depression: From Hippocratic times to modern times.* New Haven, CT: Yale University Press.

Jasanoff, A. (2018). *The biological mind.* New York: Basic Books.

Jones, E., & Wessely, S. (2005). *Shell shock to PTSD: Military psychiatry from 1900 to the Gulf War.* New York: Psychology Press.

Jones, K. D. (2012). "A critique of the DSM-5 field trials." *Journal of Nervous and Mental Disease, 200,* 517–19.

Judd, L. L., & Akiskal, H. S. (2000). "Delineating the longitudinal structure of depressive illness: Beyond thresholds and subtypes." *Pharmacopsychiatry, 33,* 3–7.

Judd, L. L., Rapaport, M. H., Paulus, M. P., & Brown, J. L. (1994). "Subsyndromal symptomatic depression: A new mood disorder?" *Journal of Clinical Psychiatry, 55,* 18–28.

Kadushin, C. (1969). *Why people go to psychiatrists.* Piscataway, NJ: AldineTransaction.

Kendell, R. E. (1983). "DSM-III: A major advance in psychiatric nosology." In *International perspectives on DSM-III,* Spitzer, R. L., Williams, J. B., & Skodol, A. E. (eds.), 55–68. Washington, DC: America Psychiatric Press, 66.

Kendell, R. E. (1991). "The relationship between the DSM-IV and the ICD-9." *Journal of Abnormal Psychology, 100,* 297–301.

Kendell, R. E., Cooper, J. E., Gourlay, A. J., Copeland, J. R., Sharpe, L., & Gurland, B. J. (1971). "Diagnostic criteria of American and British psychiatrists." *Archives of General Psychiatry, 25,* 123–30.

Kendell, R. E., & Jablensky, A. (2003). "Distinguishing between the validity and utility of psychiatric diagnoses." *American Journal of Psychiatry, 160,* 4–12.

Kendler, K. S. (1990). "Toward a scientific psychiatric nosology: Strengths and limitations." *Archives of General Psychiatry, 47,* 969–73.

Kendler, K. S. (2008). "The loss of sadness." *Psychological Medicine, 38,* 148–50.

Kendler, K., Kupfer, D., Narrow, W., Phillips, K., & Fawcett, J. (2009). "Guidelines for making changes to DSM-V." http://www.dsm5.org/ProgressReports/Pages/Default .aspx.

Kendler, K. S., Munoz, R. A., & Murphy, G. (2010). "The development of the Feighner criteria: A historical perspective." *American Journal of Psychiatry, 167,* 134–42.

Kessler, R. C., Demler, O., Frank, R. G., Olfson, M., Pincus, H. A., Walters, E. E., et al. (2005). "Prevalence and treatment of mental disorders, 1990–2003." *New England Journal of Medicine, 352,* 2515–23.

Kessler, R. C., Merikangas, K. R., Berglund, P., Eaton, W. W., Koretz, D. S., & Walters, E. E. (2003). "Mild disorders should not be eliminated from the DSM-V." *Archives of General Psychiatry, 60,* 1117–22.

Kessler, R. C., Rubinow, D. R., Holmes, C., Abelson, J. M., & Zhao, S. (1997). "The epidemiology of DSM-III-R bipolar I disorder in a general population survey." *Psychological Medicine, 27,* 1079–89.

Kessler, R. C., & Wang, P. S. (2008). "The descriptive epidemiology of commonly occurring mental disorders in the United States." *Annual Review of Public Health, 29,* 115–29.

Kiloh, L. G., Andrews, G., Neilson, M., & Bianchi, G. N. (1972). "The relationship of the syndromes called endogenous and neurotic depression." *British Journal of Psychiatry, 121,* 183–96.

Kiloh, L. G., & Garside, R. F. (1963). "The independence of neurotic depression and endogenous depression." *British Journal of Psychiatry, 109,* 451–63.

Kirk, S. A., & Kutchins, H. (1992). *The selling of DSM: The rhetoric of science in psychiatry.* New York: Aldine de Gruyter.

Klein, D. F. (1999). "Harmful dysfunction, disorder, disease, illness, and evolution." *Journal of Abnormal Psychology, 108,* 421–29.

Klein, D. F., & Fink, M. (1962). "Psychiatric reaction patterns to imipramine." *American Journal of Psychiatry, 119,* 432–38.

Kleinman, A. (1988). *Rethinking psychiatry: From cultural category to personal experience.* New York: Free Press.

Klerman, G. L. (1978). "The evolution of a scientific nosology." In *Schizophrenia: Science and practice,* Shershow, J. C. (ed.), 99–121. Cambridge, MA: Harvard University Press.

Klerman, G. L. (1983). "The efficacy of psychotherapy as the basis for public policy." *American Psychologist, 38,* 929–34.

Klerman, G. L. (1984). "The advantages of DSM-III." *American Journal of Psychiatry, 141,* 539–42.

Kraemer, H. C., Kupfer, D. J., Clarke, D. E., Narrow, W. E., & Regier, D. A. (2012a). "DSM-5: How reliable is reliable enough?" *American Journal of Psychiatry, 169*, 13–15.

Kraemer, H. C., Kupfer, D. J., Clarke, D. E., Narrow, W. E., & Regier, D. A. (2012b). "Response to Spitzer et al. letter." *American Journal of Psychiatry, 169*, 538–39.

Kraepelin, E. (1896). *Psychiatrie: Ein Lehrbuch fur Studirende und Aertze* (5th ed.). Leipzig: J. A. Barth.

Kraepelin, E. (1899). *Psychiatrie: Ein Lehrbuch fur Studirende und Aertze* (6th ed.). Leipzig: J. A. Barth.

Kramer, M. (1953). "Long range studies of mental hospital patients, an important area for research in chronic disease." *Milbank Memorial Fund Quarterly, 31*, 253–64.

Kramer, M. (1968). "The history of the efforts to agree on an international classification of mental disorders." In APA, 1968, xi–xx.

Kramer, M. (1977). *Psychiatric services and the changing institutional scene, 1950–1985.* National Institute of Mental Health, Analytical and Special Study Reports. Series B, no. 12. Washington, DC: US Government Printing Office.

Kramer, P. D. (1993). *Listening to Prozac: A psychiatrist explores antidepressant drugs and the remaking of the self.* New York: Viking.

Kramer, P. D. (2005). *Against depression.* New York: Penguin.

Krueger, R. F. (1999). "The structure of common mental disorders." *Archives of General Psychiatry, 56*, 921–26.

Kupfer, D. J., First, M. F., & Regier, D. A. (2002). *A research agenda for DSM-V.* Washington, DC: American Psychiatric Association.

Kupfer, D. J., Kuhl, E. A., & Regier, D. A. (2013). "*DSM-5*—the future arrived." *JAMA, 309*, 1691–92.

Laing, R. D. (1967). *The politics of experience and the bird of paradise.* New York: Pantheon.

Lane, C. (2006). "How shyness became an illness: A brief history of social phobia." *Common Knowledge, 12*, 388–409.

Lane, C. (2007). *Shyness: How normal behavior became a sickness.* New Haven, CT: Yale University Press.

Leaf, P. J., Myers, J. K., & McEvoy, L. T. (1991). "Procedures used in the Epidemiological Catchment Area Study." In *Psychiatric disorders in America*, Robins, L., & Regier, D. (eds.), 11–32. New York: Free Press.

Ledford, H. (2009). "Psychiatry manual revisions spark row." *Nature, 460*, 445.

LeDoux, J. (1998). *The emotional brain.* New York: Simon & Schuster.

Lehmann, H. E. (1959). "Psychiatric concepts of depression: Nomenclature and classification." *Canadian Psychiatric Association Journal, 4*, S1–S12.

Lenzer, J. (2004). "Bush plans to screen whole US population for mental illness." *British Medical Journal, 328*, 1458.

Lerner, P., & Micale, M. (2001). "Trauma, psychiatry, and history: A conceptual and historiographical introduction." In *Traumatic pasts: History, psychiatry, and trauma in the modern age, 1870–1930.* Micale, M. S., & Lerner, P. (eds.), 1–30. New York: Cambridge University Press.

Lewis, A. (1934). "Melancholia: A clinical survey of depressive states." *Journal of Mental Science, 80*, 1–43.

Lewis, A. (1967). "Melancholia: A historical review." In *The state of psychiatry: Essays and addresses*, 71–110. London: Routledge & Kegan Paul.

Lewis, A. (1979). "Classification and diagnosis in psychiatry: A historical note." In *The later papers of Sir Aubrey Lewis*, 192–210. Oxford: Oxford University Press, 1979.

Lieberman, J. A. (2012). "The truth about *DSM-V*." *Fox News*, November 15. https://www.foxnews.com/opinion/the-truth-about-dsm-5.

Lieberman, J. A. (2015). *Shrinks: The untold story of psychiatry*. New York: Little, Brown.

Lord, C., & Bishop, S. L. (2015). "Recent advances in autism research as reflected in *DSM-5* criteria for autism spectrum disorder." *Annual Review of Clinical Psychology, 11*, 53–70.

Lunbeck, E. (2014). *The Americanization of narcissism*. Cambridge, MA: Harvard University Press.

MacKinnon, C. (1988). *Feminism unmodified: Discourses on life and law*. Cambridge, MA: Harvard University Press.

Magee, W. J., Eaton, W. W., Wittchen, H. U., McGonagle, K. A., & Kessler, R. C. (1996). "Agoraphobia, simple phobia, and social phobia in the National Comorbidity Survey." *Archives of General Psychiatry, 53*, 59–68.

Martino, D. J., Szmulewicz, A. G., Valerio, M. P., & Parker, G. (2019). "Melancholia: An attempt at definition based on a review of empirical data." *Journal of Nervous and Mental Disease, 207*, 792–98.

Maser, J. D., Norman, S. B., Zisook, S., Everall, I. P., Stein, M. B., Schettler, P. J., & Judd, L. L. (2009). "Psychiatric nosology is ready for a paradigm shift in DSM-V." *Clinical Psychology Science and Practice, 16*, 24–40.

Maxmen, J. (1985). *The new psychiatrists*. New York: New American Library.

Mayes, R., & Horwitz, A. V. (2005). "DSM-III and the revolution in the classification of mental illness." *Journal of the History of Behavioral Sciences, 41*, 249–67.

McGlashan, T. H., Grilo, C. M., Skodol, A. E., et al. (2000). "The Collaborative Longitudinal Personality Disorders Study." *Acta Psychiatrica Scandinavia, 102*, 256–64.

McHugh, P. (1999). "How psychiatry lost its way." *Commentary*, December, 32–38.

McNally, R. J. (2003). *Remembering trauma*. Cambridge, MA: Harvard University Press.

McNally, R. J. (2011). *What is mental illness?* Cambridge, MA: Harvard University Press.

McNally, R. J., Bryant, R. A., & Ehlers, A. (2003). "Does early psychological intervention promote recovery from posttraumatic stress?" *Psychological Science in the Public Interest, 4*, 45–79.

McPartland, J. C., Reichow, B., & Volkmar, F. R. (2012). "Sensitivity and specificity of proposed DSM-5 diagnostic criteria for autism spectrum disorder." *Journal of the Academy of Child and Adolescent Psychiatry, 51*, 368–83.

Mechanic, M. (2013). "Psychiatry's new diagnostic manual: 'Don't buy it. Don't use it. Don't teach it.'" *Mother Jones*, May 14, 1–8.

Meehl, P. E. (1972). *Psychodiagnosis: Collected papers*. New York: W. W. Norton.

Mendels, J. (1968). "Depression: The distinction between syndrome and symptom." *American Journal of Psychiatry, 114*, 1349–54.

Mendels, J., & Cochrane, C. (1968). "The nosology of depression: The endogenous-reactive concept." *American Journal of Psychiatry, 124*, 1–11.

Menninger, K. (1963). *The vital balance.* New York: Viking Press.

Menninger, W. (1945). "Neuropsychiatry for the general medical officer." *Mental Hygiene, 29,* 642–52.

Metzl, J. (2003). *Prozac on the couch: Prescribing gender in the era of wonder drugs.* Durham, NC: Duke University Press.

Mewton, L., Slade, T., McBride, O., Grove, R., & Teesson, M. (2010). "An evaluation of the proposed DSM-5 alcohol use disorder criteria using Australian national data." *Addiction, 106,* 941–50.

Meyer, A. (1948). *The commonsense psychiatry of Adolf Meyer: Fifty-two selected papers.* Lief, A. (ed.). New York: McGraw-Hill.

Miller, L. S., Bergstrom, D. A., Cross, H. J., & Grube, J. W. (1981). "Opinions and use of the *DSM* system by practicing psychologists." *Professional Psychology, 12,* 385–90.

Millon, T. (2012). "On the history and future study of personality and its disorders." *Annual Review of Clinical Psychology, 8,* 1–19.

Mirowsky, J., & Ross, C. E. (2002). "Measurement for a human science." *Journal of Health and Social Behavior, 43,* 152–70.

Mischel, W. (1968). *Personality and assessment.* Mahwah, NJ: Lawrence Erlbaum.

Mojtabai, R. (2011). "Bereavement-related depressive episodes: Characteristics, 3-year course, and implications for DSM-5." *Archives of General Psychiatry, 68:* 920–28.

Mojtabai, R., & Olfson, M. A. (2008). "National patterns in antidepressant treatment by psychiatrists and general medical providers: Results from the National Comorbidity Survey Replication." *Journal of Clinical Psychiatry, 69,* 1064–74.

Moreno, C., Laje, G., Blanco, C., Jiang, H., Schmidt, A. B., & Olfson, M. (2007). "National trends in the outpatient diagnosis and treatment of bipolar disorder in youth." *Archives of General Psychiatry, 64,* 1032–39.

Mościcki, E. K., Clarke, D. E., Kuramoto, S. J., Kraemer, H.C., Narrow, W. E., Kupfer, D. J., & Regier, D. A. (2013). "Testing DSM-5 in routine clinical practice settings: Feasibility and clinical utility." *Psychiatric Services, 64,* 952–60.

Moynihan, R., & Cassels, A. (2005). *Selling sickness: How the world's biggest pharmaceutical companies are turning us all into patients.* New York: Nation Books.

Murray, C. J. L., & Lopez, A. D. (eds.). (1996). *The global burden of disease.* Cambridge, MA: World Health Organization.

Murray, R. M. (1979). "A reappraisal of American psychiatry." *Lancet, 8110,* 255–58.

Narrow, W. E., Clarke, D. E., Kuramoto, S. J., Kraemer, H. C., Kupfer, D. J., Greiner, L., & Regier, D. A. (2013). "DSM-5 field trials in the United States and Canada, part III: Development and reliability testing of a cross-cutting symptom assessment for DSM-5." *American Journal of Psychiatry, 170,* 71–82.

Nathan, P. (1994). "DSM-IV: Empirical, accessible, not yet ideal." *Journal of Clinical Psychology, 50,* 103–10.

National Association of State Mental Health Program Directors. (2017). *Trend in psychiatric inpatient capacity, United States and each state: 1970–2014.* Alexandria: Author.

National Disease and Therapeutic Index. *Encyclopedia of Public Health.* Retrieved June 20, 2019 from *Encyclopedia.com,* https://www.encyclopedia.com/education /encyclopedias-almanacs-transcripts-and-maps/national-disease-and-therapeutic -index.

NDTI Review (1970). "The psychiatrist in private practice." *1*(1), March, 1–20.

Nesse, R. M. (2019). *Good reasons for bad feelings: Insights from the frontier of evolutionary psychiatry.* New York: Dutton.

Nesse, R. M., & Jackson, E. D. (2006). "Evolution: Psychiatric nosology's missing biological foundation." *Clinical Neuropsychiatry, 3,* 121–31.

Nestler, E. J. (2018). "New approaches for treating depression." In *Charney & Nestler's neurobiology of mental illness* (5th ed.), Charney, D. S., Buxbaum, J. D., Sklar, P., & Nestler, E. J. (eds.), 377–85. New York: Oxford University Press.

Noll, R. (2011). *American madness: The rise and fall of dementia praecox.* Cambridge, MA: Harvard University Press.

Norris, F. H., Friedman, M. J., Watson, P. J., Byrne, C. M., Diaz, E., & Kaniasty, K. (2002). "60,000 disaster victims speak: Part I. An empirical review of the empirical literature, 1981–2001." *Psychiatry, 65,* 207–39.

Ofshe, R., & Watters, E. (1994). *Making monsters: False memories, psychotherapy, and sexual hysteria.* Berkeley: University of California Press.

Olfson, M., Blaco, C., Liu, L., Moreno, C., & Laje, G. (2006). "National trends in the outpatient treatment of children and adolescents with antipsychotic drugs." *Archives of General Psychiatry, 63,* 679–85.

Olfson, M., & Klerman, G. L. (1993). "Trends in the prescription of antidepressants by office-based psychiatrists." *American Journal of Psychiatry, 150,* 571–77.

Olfson, M., Kroenke, K., Wang, S., & Blanco, C. (2014). "Trends in office-based mental health care provided by psychiatrists and primary care physicians." *Journal of Clinical Psychiatry, 75,* 247–53.

Olfson, M., Marcus, S. C., Druss, B., Elinson, L., Tanielian, T., & Pincus, H. A. (2002a). "National trends in the outpatient treatment of depression." *Journal of the American Medical Association, 287,* 203–9.

Olfson, M., Marcus, S. C., Druss, B., & Pincus. H. A. (2002b). "National trends in the use of outpatient psychotherapy." *American Journal of Psychiatry, 159,* 1914–20.

Overall, J. E., Hollister, L. E., Johnson, M., & Pennington, V. (1966). "Nosology of depression and differential response to drugs." *Journal of the American Medical Association, 195,* 162–64.

Paris, J. (2015). *Overdiagnosis in psychiatry.* New York: Oxford University Press.

Parker, G., Fink, M., Shorter, E., Taylor, M. A., Akiskal, H., Berrios, G., et al. (2010). "Issues for DSM-5: Whither melancholia? The case for its classification as a distinct mood disorder." *American Journal of Psychiatry, 167,* 745–47.

Parker, G., McClure, G., & Paterson, A. (2015). "Melancholia and catatonia: Disorders or specifiers?" *Current Psychiatry Reports, 17,* 536.

Pasamanick, B., Dinitz, S., & Lefton, M. (1959). "Psychiatric orientation and its relation to diagnosis and treatment in a mental hospital." *American Journal of Psychiatry, 116,* 127–32.

Paton, S. (1905). *Psychiatry: A text-book for students and physicians.* Philadelphia: J. B. Lippincott.

Paul, A. M. (2004). *The cult of personality testing.* New York: Free Press.

Paykel, E. S. (1971). "Classification of depressed patients: A cluster analysis derived grouping." *British Journal of Psychiatry, 118,* 275–88.

Pettersson, E., Larsson, H., & Lichtenstein, P. (2016). "Common psychiatric disorders share the same genetic origin: A multivariate sibling study of the Swedish population." *Molecular Psychiatry, 21,* 717–21.

Phalen, P. L., Rouhakhtar, P. R., Millman, Z. B., Thompson, E., DeVylder, J., Mittal, V., et al. (2018). "Validity of a two-item screen for early psychosis." *Psychiatry Research, 270,* 861–68.

Phillips, J. (2011). "The great DSM-5 personality bazaar." *Psychiatric Times.* https://www.psychiatrictimes.com/view/great-dsm-5-personality-bazaar.

Pies, R. (2008). "Major depression after recent loss is major depression—until proven otherwise." *Psychiatric Times,* December 12, 12.

Pilecki B. C., Clegg, J. W., & McKay D. (2011) "The influence of corporate and political interests on models of illness in the evolution of the DSM." *European Psychiatry, 26,* 194–200.

Pilkonis, P. A., Hallquist, M. N., Morse, J. Q., & Stepp, S. D. (2011). "Striking the (im)proper balance between scientific advances and clinical utility: Commentary on the DSM-5 proposal for personality disorders." *Personality Disorders, 2,* 68–82.

Pollock, B. (1959). "Clinical findings in the use of Tofranil in depressive and other psychiatric states." *American Journal of Psychiatry, 116,* 312–17.

Pols, H. (2001). "Divergences in American psychiatry during the Depression: Somatic psychiatry, community mental hygiene, and social reconstruction." *Journal of the History of the Behavioral Sciences, 37,* 369–88.

Porter, R., & Micale, M. S. (1994). "Introduction: Reflections on psychiatry and its histories." In *Discovering the history of psychiatry,* Micale, M. S., & Porter, R. (eds.), 348–83. New York: Oxford University Press.

Pressman, J. D. (2002). *Last resort: Psychosurgery and the limits of medicine.* Cambridge: Cambridge University Press.

Raines, G. N. (1952). "Foreword." In APA, 1952, v–xi.

Raines, G. N. (1953). "The new nomenclature." *American Journal of Psychiatry, 109,* 548–59.

Raofi, S., & Schappert, S. M. (2006). "Medication therapy in ambulatory medical care, United States 2003–04." *Vital Health Statistics, 13,* 1–40.

Raskin, A., & Crook, T. H. (1976). "The endogenous-neurotic distinction as a predictor of response to antidepressant drugs." *Psychological Medicine, 6,* 59–70.

Ray, I. (1938/1962). *A treatise on the medical jurisprudence of insanity.* Cambridge, MA: Harvard University Press.

Raynes, N. (1979). "Factors affecting the prescribing of psychotropic drugs in general practice consultations." *Psychological Medicine, 9,* 671–79.

Regier, D. A., Kuhl, E. A., & Kupfer, D. J. (2013). "The DSM-5: Classification and criteria changes." *World Psychiatry, 12,* 92–98.

Regier, D. A., Narrow, W. E., Clarke, D. E., Kraemer, H. C., Kuramoto, S. J., Kuhl, E. A., & Kupfer, D. J. (2013). "DSM-5 field trials in the United States and Canada, part II: Test-retest reliability of selected categorical diagnoses." *American Journal of Psychiatry, 170,* 59–70.

Regier, D. A., Narrow, W. E., Kuhl, E. A., & Kupfer, D. J. (2009). "The conceptual development of *DSM-V.*" *American Journal of Psychiatry, 166,* 983–87.

Regier, D. A., Narrow, W. E., Kuhl, E. A., & Kupfer, D. J. (eds.). (2011). *The conceptual evolution of DSM-5*. Washington, DC: American Psychiatric Association Press.

Regier, D. A., Narrow, W. E., & Rae, D. S. (2004). "For DSM-V, it's the 'disorder threshold,' stupid." *Archives of General Psychiatry, 61*, 1051–52.

Rickels, K. R., & Rynn, M. A. (2001). "What is generalized anxiety disorder?" *Journal of Clinical Psychiatry, 62*, 4–12.

Rickles, N. K., Klein, J. J., & Bassan, M. E. (1950). "Who goes to a psychiatrist?" *American Journal of Psychiatry, 106*, 845–50.

Rieff, P. (1966). *The triumph of the therapeutic*. Chicago: University of Chicago Press.

Robins, E., & Guze, S. B. (1970). "Establishment of diagnostic validity in psychiatric illness: Its application to schizophrenia." *American Journal of Psychiatry, 126*, 983–97.

Robins, L. N., & Helzer, J. E. (1986). "Diagnosis and clinical assessment: The current state of psychiatric diagnosis." *Annual Review of Psychology, 37*, 409–32.

Robins, L. N., & Regier, D. A. (eds.). (1991). *Psychiatric disorders in America: The Epidemiological Catchment Area Study*. New York: Free Press.

Ropper, A. H., & Burrell, B. D. (2019). *How the brain lost its mind: Sex, hysteria, and the riddle of mental illness*. New York: Avery.

Rose, N. (2006). *The politics of life itself*. Princeton, NJ: Princeton University Press.

Rosenberg, C. (2007). *Our present complaint: American medicine, then and now*. Baltimore: Johns Hopkins University Press.

Rosenhan, D. (1973). "On being sane in insane places." *Science, 179*, 250–58.

Roth, M. (1990). "Categorical and unitary classifications of neurotic disorder." *Journal of the Royal Society of Medicine, 83*, 609–14.

Ruscio, A. M., Chiu, W. T., Roy-Byrne, P., Stang, P. E., Stein, D. J., Wittchen, H.-U., & Kessler, R. C. (2007). "Broadening the definition of generalized anxiety disorder: Effects on prevalence and associations with other disorders in the National Comorbidity Survey Replication." *Journal of Anxiety Disorders, 21*, 662–76.

Sabshin, M. (1990). "Turning points in twentieth-century American psychiatry." *American Journal of Psychiatry, 147*, 1267–74.

Sadler, J. Z. (2005). *Values and psychiatric diagnosis*. New York: Oxford University Press.

Satel, S. (2010). "The battle over battle fatigue." *Wall Street Journal*, July 17. https://www.wsj.com/articles/SB10001424052748704913304575371130876271708.

Schappert, S. M., & Rechtsteiner, E. A. (2008). "Ambulatory medical care utilization estimates for 2006." *National Health Statistics Reports, 6*, 1–29.

Schatzberg, A. F., Scully, J. H., Kupfer, D. J., & Regier, D. A. (2009). "Setting the record straight." *Psychiatric Times*, July 1. https://www.psychiatrictimes.com/view/setting-record-straight-response-frances-commentary-dsm-v.

Schnittker, J. (2017). *The diagnostic system: Why the classification of psychiatric disorders is necessary, difficult, and never settled*. New York: Columbia University Press.

Schwartz, M. A., & Wiggins, O. P. (2002). "The hegemony of the DSMs." In *Descriptions and prescriptions values, mental disorders, and the DSMs*, Sadler, J. (ed.), 199–209. Baltimore: Johns Hopkins University Press.

Scull, A. (2005). *Madhouse: A tragic tale of megalomania and modern medicine*. New Haven, CT: Yale University Press.

Scull, A. (2019). *Psychiatry and its discontents.* Berkeley: University of California Press.

Sharfstein, S. S. (2005). "Big pharma and American psychiatry: The good, the bad, and the ugly." *Psychiatric News,* August 19. https://psychnews.psychiatryonline.org/doi /full/10.1176/pn.40.16.00400003.

Shatan, C. (1972). "Post-Vietnam syndrome." *New York Times,* May 6.

Shorter, E. (2009). *Before Prozac: The troubled history of mood disorders in psychiatry.* New York: Oxford University Press.

Shorter, E. (2013). *How everyone became depressed: The rise and fall of the nervous breakdown.* New York: Oxford University Press.

Shorter, E. (2015). *What psychiatry left out of the DSM-5: Historical mental disorders today.* New York: Routledge.

Simon, B. (1978). *Mind and madness in Ancient Greece.* Ithaca, NY: Cornell University Press.

Simons, J. (2004). "Lilly goes off Prozac: The drugmaker bounced back from the loss of its blockbuster, but the recovery had costs." *Fortune,* June 28.

Skodol, A. E. (2000). "Diagnosis and classification of mental disorders." In *American psychiatry after World War II: 1944–1994,* Menninger, R.W., & Nemiah, J. C. (eds.), 430–58. Washington, DC: American Psychiatric Press.

Skodol, A. E. (2012). "Personality disorders in DSM-5." *Annual Review of Clinical Psychology, 8,* 317–44.

Slovenko, R. (2011). "The DSM in litigation and legislation." *Journal of the American Academy of Psychiatry and Law, 39,* 6–11.

Smith, D. T. (2019). *Medicine over mind: Mental health practice in the biomedical era.* New Brunswick, NJ: Rutgers University Press.

Smith, M. B. (1985). *Small comfort: A history of the minor tranquillizers.* New York: Praeger.

Sommers, C. H., & Satel, S. (2006). *One nation under therapy.* New York: St. Martins Griffin.

Spiegel, A. (2005). "The dictionary of disorder: How one man revolutionized psychiatry." *New Yorker,* January 3, 56–63.

Spitzer, R. L. (1975). "On pseudoscience in science, logic in remission and psychiatric diagnosis: A critique of Rosenhan's 'On being sane in insane places.'" *Journal of Abnormal Psychology, 84,* 442–52.

Spitzer, R. L. (1982) "Letter to the editor." *Schizophrenia Bulletin, 8,* 592.

Spitzer, R. L. (1991). "An outsider-insider's views about revising the DSMs." *Journal of Abnormal Psychology, 100,* 294–96.

Spitzer, R. L. (1999). "Harmful dysfunction and the DSM-III definition of mental disorder." *Journal of Abnormal Psychology, 108,* 430–32.

Spitzer, R. L., Endicott, J., & Robins, E. (1975). "Clinical criteria for psychiatric diagnosis and DSM-III." *American Journal of Psychiatry, 132,* 1187–92.

Spitzer, R. L., Endicott, J., & Robins, E. (1978). "Research diagnostic criteria: Rationale and reliability." *Archives of General Psychiatry, 35,* 773–82.

Spitzer, R. L., & Fleiss, J. L. (1974). "A re-analysis of the reliability of psychiatric diagnosis." *American Journal of Psychiatry, 125,* 341–47.

Spitzer, R. L., Severino, S., Williams, J. B. W., & Perry, B. (1989). "Late luteal phase dysphoric disorder and DSM-III-R." *American Journal of Psychiatry, 146,* 892–97.

Spitzer, R. L., Sheehy, M., & Endicott, J. (1977). "DSM-III: Guiding principles." In *Psychiatric Diagnosis*, Rakoff, V., Stancer, H., & Kedward, H. (eds.), 1–24. New York: Bruner/Mazel.

Spitzer, R. L., & Williams, J. B. W. (1983). "Classification in psychiatry." In *Comprehensive textbook of psychiatry*, Kaplan, H. I., & Sadock, B. J. (eds.), 591–613. Baltimore: Williams & Wilkins.

Spitzer, R. L., Williams, J. B. W., & Endicott, J. (2012). "Standards for DSM-5 reliability." *American Journal of Psychiatry, 169*, 537.

Spitzer, R. L., & Wilson, P. T. (1968). "A guide to the new nomenclature." In APA, 1968, 120–34.

Sprooten, E., Rasgon, A., Goodman, M., Carlin, A., Leibu, E., Lee, W. H., & Frangou, S. (2017). "Addressing reverse inference in psychiatric neuroimaging: Meta-analyses of task-related brain activation in common mental disorders." *Human Brain Mapping, 38*, 1846–64.

Starcevic, V., & Portman, M. E. (2013). "The status quo as a good outcome: How the DSM-5 diagnostic criteria for generalized anxiety disorder remained unchanged from the DSM-IV criteria." *Australian and New Zealand Journal of Psychiatry, 47*, 995–97.

Starcevic, V., Portman, M. E., & Beck, A. T. (2012). "Generalized anxiety disorder: Between neglect and an epidemic." *Journal of Nervous and Mental Disease, 200*, 664–67.

Statistical manual for the use of institutions for the insane prepared by the Committee on Statistics of the American Medico-Psychological Association in collaboration with the Bureau of Statistics of the National Committee for Mental Hygiene. (1918). New York: National Committee for Mental Hygiene.

Staub, M. E. (2011). *Madness is civilization: When the diagnosis was social, 1948–1980*. Chicago: University of Chicago Press.

Stein, M. B., Walker, J. R., & Forde, D. R. (1994). "Setting diagnostic thresholds for social phobia: Considerations from a community survey of social anxiety." *American Journal of Psychiatry, 151*, 408–12.

Steinem, G. (1983). *Outrageous acts and everyday rebellions*. New York: Henry Holt.

Summerfield, D. (2001). "The invention of post-traumatic stress disorder and the social usefulness of a psychiatric category." *British Medical Journal, 322*, 95.

Szasz, T. (1961). *The myth of mental illness: Foundations of a theory of personal conduct*. New York: Harper & Row.

Talbott, J. (1980). "An in-depth look at DSM-III: An interview with Robert Spitzer." *Hospital and Community Psychiatry, 31*, 25–32.

Tone, A. (2009). *The age of anxiety: A history of America's turbulent affair with tranquilizers*. New York: Basic Books.

Travis, C. (1993). "You haven't come very far, baby." *Los Angeles Times*, March 4, B7.

Trull, T. J., & Widiger, T. A. (2013). "Dimensional models of personality: The five-factor model and the DSM-5." *Dialogues in Clinical Neuroscience, 15*, 135–46.

Tuke, D. H. (1892). *A dictionary of psychological medicine*. Vol. 1. London: Churchill.

Tyrer, P. (1984). "Classifications of anxiety." *British Journal of Psychiatry, 144*, 77–83.

US Department of Health and Human Services (USDHHS). 1999. *Mental health: A report of the surgeon general.* Rockville, MD: Author.

Van Dam, N. T., Iacoviello, B. M., & Murrough, J. W. (2018). "Diagnosis and epidemiology of depression." In *Charney & Nestler's neurobiology of mental illness* (5th ed.), Charney, D. S., Buxbaum, J. D., Sklar, P., & Nestler, E. J. (eds.), 289–300. New York: Oxford University Press.

van Praag, H. M. (1990). "The DSM-IV (depression) classification: To be or not to be?" *Journal of Nervous and Mental Disease, 178,* 147–49.

Vedantam, S. (2001). "Drug ads hyping anxiety make some uneasy." *Washington Post,* July 16. https://www.washingtonpost.com/archive/politics/2001/07/16/drug-ads-hyping-anxiety-make-some-uneasy/8fe2eea2-b780-48cd-9872-1d3802e83147/.

Verhoeff, B. (2010). "Drawing borders of mental disorders: An interview with David Kupfer." *Biosocieties, 5,* 467–75.

Veroff, J., Kulka, R. A., & Douvan, E. (1981). *Mental health in America.* New York: Harper & Row.

Visser, S. N., Danielson, M. L., Bitsko, R. H., Holbrook, J. R., Kogan, M. D., Ghandour, R. M., et al. (2014). "Trends in the parent-report of health care provider diagnosed and medicated ADHD: United States, 2003–2011." *Journal of the American Academy of Child and Adolescent Psychiatry, 53,* 34–46.

Wakefield, J. C. (1992). "The concept of mental disorder: On the boundary between biological facts and social values." *American Psychologist, 47,* 373–88.

Wakefield, J. C. (1996). "DSM-IV: Are we making diagnostic progress?" *Contemporary Psychology, 41,* 646–52.

Wakefield, J. C. (2001). "The myth of DSM's invention of new categories of disorder: Houts's diagnostic discontinuity thesis disconfirmed." *Behavior Research and Therapy, 39,* 575–624.

Wakefield, J. C. (2006). "Diagnostic issues and controversies in DSM-5: Return of the false positives problem." *Annual Review of Clinical Psychology, 12,* 105–32.

Wakefield, J. C. (2015a). "DSM-5, psychiatric epidemiology and the false positives problem." *Epidemiology and Psychiatric Science, 24,* 188–96.

Wakefield, J. C. (2015b). "DSM-5 substance use disorder: How conceptual missteps weakened the foundations of the addictive disorders field." *Acta Psychiatrica Scandinavica, 132,* 327–34.

Wakefield, J. C., & First, M. B. (2012). "Placing symptoms in context: The role of contextual criteria in reducing false positives in diagnostic and statistical manual of mental disorders diagnoses." *Comprehensive Psychiatry, 53,* 130–39.

Wakefield, J. C., & Horwitz, A.V. (2016). "Psychiatry's continuing expansion of depressive disorder." In *Sadness or depression?* Demazeux, S. (ed), 173–203. Dordrecht: Springer.

Wakefield, J. C., & Schmitz, M. F. (2012). "Recurrence of depression after bereavement-related depression: Evidence for the validity of the DSM-IV bereavement exclusion from the Epidemiologic Catchment Area Study." *Journal of Nervous and Mental Disease, 200,* 480–85.

Wakefield, J. C., & Schmitz, M. F. (2013a). "Normal vs. disordered bereavement-related depression: Are the differences real or tautological?" *Acta Psychiatrica Scandinavica, 127,* 159–68.

Wakefield, J. C., & Schmitz, M. F. (2013b). "Can the DSM's major depression bereavement exclusion be validly extended to other stressors? Evidence from the NCS." *Acta Psychiatrica Scandinavica, 128,* 294–305.

Wakefield, J. C., Schmitz, M. F., First, M. B., & Horwitz, A. V. (2007). "Extending the bereavement exclusion for major depression to other losses: Evidence from the National Comorbidity Survey." *Archives of General Psychiatry, 64,* 433–40.

Waldon, E. G. (2014). "DSM-5: Changes and controversies." *Music Therapy Perspectives,* 32(1), 78–83. https://doi.org/10.1093/mtp/miu011. Published: 26 June 2014.

Walker, J. (1993). *Couching resistance: Women, film, and psychoanalytic psychiatry.* Minneapolis: University of Minnesota Press.

Wallace, E. R. (1994). "Psychiatry and its nosology: A historico-philosophical overview." In *Philosophical perspectives on psychiatric diagnostic classification*, Sadler, J. Z., Wiggins, O. P., & Schwartz, M. A. (eds), 6–86. Baltimore: Johns Hopkins University Press.

Ward, S. C. (2002). *Modernizing the mind: Psychological knowledge and the remaking of society.* Westport, CT: Prager.

War Department Technical Bulletin. Medical 203. (1946). "Nomenclature of psychiatric disorders and reactions." *Journal of Clinical Psychology, 2,* 289–96.

Weiden, P. J. (2012). "The risk that DSM-5 will promote even more inappropriate antipsychotic exposure in children and teenagers." *Current Psychiatry Reviews, 8,* 271–76.

Weiner, D. B. (1994). "*Le geste de Pinel*: The history of a psychiatric myth." In *Discovering the history of psychiatry*, Micale, M. S., and Porter, R. (eds), 232–47. New York: Oxford University Press.

Weissman, M. M., Wickramaratne, P., Nomura, Y., Warner, V., Pilowsky, D., & Verdeli, H. (2006). "Offspring of depressed parents: 20 years later." *American Journal of Psychiatry, 163,* 1001–8.

Wessely, S. (2005). "Victimhood and resilience." *New England Journal of Medicine, 353,* 548–50.

Whiteside, T. (1958). "Marketing Miltown." *New Yorker,* May 5, 115–20.

Whooley, O. (2014). "Nosological reflections: The failure of DSM-5, the emergence of RDoC, and the decontextualization of mental distress." *Society and Mental Health, 4,* 92–110.

Whooley, O. (2016). "Measuring mental disorders: The failed commensuration project of DSM-5." *Social Science and Medicine, 66,* 33–40.

Whooley, O. (2019). *On the heels of ignorance: Psychiatry and the politics of not knowing.* Chicago: University of Chicago Press.

Whooley, O., & Horwitz, A. V. (2013). "The paradox of professional success: Grand ambition, furious resistance, and the derailment of the DSM-5 revision." In *Making the DSM-5: Concepts and controversies*, Paris, J., & Phillips, J. (eds), 75–94. New York: Springer.

Williams, S. J., Martin, P., & Gabe, J. (2011). "The pharmaceuticalisation of society? A framework for analysis." *Sociology of Health and Illness, 33,* 710–25.

Wilson, M. (1993). "DSM-III and the transformation of American psychiatry: A history." *American Journal of Psychiatry, 150,* 399–410.

Winokur, G., & Clayton, P. (1967). "Family history studies: I. Two types of affective disorders separated according to genetic and clinical factors." In *Recent advances in biological psychiatry*, Wortis, J. (ed.), 25–30. New York: Plenum.

Wong, A., Choy, H., & Van Tol, H. H. M. (2003). "Schizophrenia: From phenomenology to neurobiology." *Neuroscience and Biobehavioral Reviews, 27,* 269–306.

World Health Organization. (1979). *International classification of diseases* (9th rev.). Salt Lake City: Medicode.

World Health Organization. (1992). *The ICD-10 classification of mental and behavioural disorders: Clinical descriptions and diagnostic guidelines.* Geneva: Author.

World Health Organization. (2018). *International classification of diseases for mortality and morbidity statistics* (11th rev.). https://icd.who.int/browse11/l-m/en.

Zachar, P., First, M. B., & Kendler, K. S. (2017). "The bereavement exclusion debate in the DSM-5: A history." *Clinical Psychological Science, 5,* 890–906.

Zachar, P., First, M. B., & Kendler, K. S. (2020). "The DSM-5 proposal for attenuated psychosis syndrome: A history." *Psychological Medicine, 50,* 1–7.

Zachar, P., & Kendler, K. S. (2014). "A Diagnostic and statistical manual of mental disorders history of premenstrual dysphoric disorder." *Journal of Nervous and Mental Disorders, 202,* 346–52.

Zachar, P., & Krueger, R. F. (2013). "Personality disorder and validity: A history of the controversy." In *The Oxford handbook of philosophy and psychiatry*, Fulford, K. W. M., et al. (eds.), 889–910. New York: Oxford University Press.

Zachar, P., Krueger, R. F., & Kendler, K. S. (2016). "Personality disorder in DSM-5: An oral history." *Psychological Medicine, 46,* 1–10.

Zachar, P., Regier, D. A., & Kendler, K. S. (2019). "The aspirations for a paradigm shift in DSM-5: An oral history." *Journal of Nervous and Mental Disease, 207,* 778–84.

Zerubavel, E. (1991). *The fine line.* Chicago: University of Chicago Press.

Zerubavel, E. (1999). *Social mindscapes.* Cambridge, MA: Harvard University Press.

Zuvekas, S. H. (2005). "Prescription drugs and the changing patterns of treatment for mental disorders, 1996–2001." *Health Affairs, 24,* 195–205.

Index

Caplan, Paula, 7, 92, 93
Carpenter, William, 127
child and adolescent disorders: in *DSM-I*, 152; in
    DSM-II, 31, 32, 42, 112, 152, 153, 168n87; in
    DSM-III, 82–83, 112, 153; in *DSM-III-R*, 153; in
    DSM-IV, 12, 153; in *DSM-5*, 139–40, 153, 155;
    expansion of pathology for, 151–56; as major
    class, 10, 12; medications, 12, 112–13, 151; and
    mental hygiene movement, 126; need for
    specific diagnoses, 10, 12, 139, 140, 150, 163;
    screenings, 126. *See also* autism; psychosis risk
    syndrome
childhood bipolar disorder (CBD), 112–13, 151, 155
clinicians *vs.* researchers: and depression, 68; and
    dimensional approach, 129–32; and *DSM-II*, 31,
    146; and *DSM-III*, 49, 51–53, 55, 58–59, 84,
    85–86, 115, 146–47; and *DSM-5*, 117–20,
    129–32, 142–43, 147–49, 162–63; *DSM* uses by,
    9, 84, 115; and need for specific diagnoses,
    49–50, 146, 163; in overview, 9, 10; and
    personality disorders, 11, 76–77, 148
Compendium of Psychiatry (Kraepelin), 17–18
Costello, Jane, 182n55
Cullen, William, 166n3
culture: anti-psychiatry movement, 74, 158; and
    bipolar disorders, 112, 113; contexts of, 159,
    161–62; and depression, 173n17; and *DSM-I*, 3,
    29–30, 33, 35, 40, 88, 145–46; and *DSM-II*, 3,
    33, 34, 35, 40, 88, 149; and *DSM-III*, 83, 85,
    88–89, 116, 149–50; and *DSM-III-R/DSM-IV*,
    88–89, 99; and *DSM-5*, 88–89, 99, 161–62;
    general influences, 1, 144–50; perceptions of
    psychiatry, 24, 33–34, 49, 51; and personality
    disorders, 1, 29–30, 92, 94–95; and PTSD, 99,
    149–50, 154; and tranquilizers, 36–37

dementia praecox. *See* schizophrenia
dependence. *See* substance use disorders
dependent personality disorder, 78, 131
depression: and culture, 173n17; in *DSM-I*, 28–29,
    67, 73, 133, 146, 160; in *DSM-II*, 32, 67, 68, 73,
    133, 146, 160; in *DSM-III*, 57–58, 67–72, 73, 75,
    101, 146, 160; in *DSM-III-R*, 101, 160; in
    DSM-IV, 101, 160, 178n43, 179n98; in *DSM-5*,
    124, 133–36, 155, 160; in Feighner criteria,
    170n29; historical approaches to, 11, 15, 16, 17,

67, 100–101; in *ICD*, 73; in Kraepelin system,
    18, 57, 65, 75, 109; as major class, 10; in
    Medical 203, 22, 167n39; medications, 45, 68,
    72, 100–106, 133; mixed with anxiety, 54, 73,
    124, 179n98; and PTSD, 80; rates of, 34, 68, 71,
    101–2, 135, 146; in RDC, 54; reliability study,
    74; replacement of anxiety with, 71–72, 73,
    102–4, 106; terms for, 67. *See also* bipolar
    disorders; major depressive disorder
developmental disorders, 66, 83, 139, 140, 153
diagnoses: challenges of, 2, 3, 6–7, 16, 119–20;
    feminization of, 90–93; historical approaches
    to, 1–2, 14–25; multiple, 42, 66; numbers of,
    31, 59, 66–67, 87, 152; and social control, 6, 7,
    47, 78, 92. *See also* reliability; validity
diagnoses, need for specific: advocacy groups, 9,
    10, 94–99, 158; disinterest in, 2, 8, 15, 17–19,
    24–25, 26, 30–35, 39–40, 44, 51, 146; drug
    industry, 3, 9, 10, 44, 49, 52, 72, 99–114, 145,
    163; and *DSM-III*, 44–51; general interest, 2, 8,
    10; government, 3, 4–5, 8, 9, 10, 44, 46–49, 52,
    85, 145, 149, 150, 163; individuals, 9–10, 12, 81,
    139, 145, 149, 150, 151, 163; insurers, 8, 10, 44,
    49–50, 53, 58, 70, 145, 146, 150, 163
*Diagnostic and Statistical Manual of Mental
    Disorders. See DSM*
dimensional approaches, 23, 128–32, 139–40,
    142–43, 148–49
disruptive mood dysregulation disorder (DMDD),
    124, 155
dissociation, 22, 28, 61, 94, 95–96
dissociative identity disorder (DID), 95–96.
    See also multiple personality disorder
distress: and homosexuality, 41, 79, 114; in
    mental disorders, 62–63, 159–60, 161, 175n75;
    in PTSD, 97; and social phobia, 106
domestic abuse, 92, 96
drug abuse. *See* substance use disorders
drug industry: ads, 9, 36–38, 47–48, 103–9,
    111–12, 147, 151; and child disorders, 12, 112–13,
    127, 155; and depression, 68, 72, 102–6; and
    DSM-I/DSM-II, 35–40; and *DSM-III*, 49, 52, 72,
    85, 114–15; and *DSM-III-R/DSM-IV*, 89, 90,
    99–114; and *DSM-5*, 122, 123, 151; financial ties
    with, 9, 122, 123, 151, 155; need for specific
    diagnoses, 3, 9, 10, 44, 49, 52, 72, 99–114, 145,